Self Aware Security for Real Time Task Schedules in Reconfigurable Hardware Platforms

Krishnendu Guha · Sangeet Saha ·
Amlan Chakrabarti

Self Aware Security for Real Time Task Schedules in Reconfigurable Hardware Platforms

 Springer

Krishnendu Guha
A. K. Choudhury School of Information
Technology (AKCSIT)
University of Calcutta
Kolkata, West Bengal, India

Sangeet Saha
School of Computer Science and Electronic
Engineering (CSEE)
University of Essex
Colchester, UK

Amlan Chakrabarti
A. K. Choudhury School of Information
Technology (AKCSIT)
University of Calcutta
Kolkata, West Bengal, India

ISBN 978-3-030-79703-4   ISBN 978-3-030-79701-0 (eBook)
https://doi.org/10.1007/978-3-030-79701-0

This Springer imprint is published by the registered company Springer Nature Switzerland AG
The registered company address is: Gewerbestrasse 11, 6330 Cham, Switzerland

To our Families

Preface

Security is of prime importance since times immemorial. However, the nature of security changes with the demand of industry. Industry 1.0 was associated with mechanization, steam power, weaving machines, etc., where the security objective was to prevent mechanical damage of system components. Industry 2.0 witnessed the use of electrical energy for mass production and distribution of products. Along with the security needs of Industry 1.0, the concept of securing the physical supply chain was a key aspect at this time. With Industry 3.0, the digital era started, where electronics and information technology were used to enhance automation. Though systems were automated, yet human input and intervention were needed. Security from intentional attacks gained importance in this era. However, the key focus was associated with software security and hardware was considered trusted. Presently, we are in Industry 4.0, which is the era of smart machines. This era features cyber-physical systems (CPSs) and internet of things (IoTs). With the aid of artificial intelligence (AI) and machine learning (ML), devices can communicate and interact with each other, trigger actions and even control one another without the aid of human support. To enhance speed of operations and eliminate attacks related to software, direct task execution on hardware was seen. However, on the flip side, the era has witnessed hardware threats. Thus, the security need in this domain is not only confined to software and network security but also hardware security.

Moreover, this era has witnessed the advent of field-programmable gate arrays (FPGAs) that are more flexible and replaces the traditional application-specific integrated circuits (ASICs), which are rigid. The property of dynamic partial reconfiguration of FPGAs facilitate spatial and temporal scheduling and execution of various tasks in the same platform. Thus, FPGAs find usage in a wide range of applications from simple smart home appliances to complex nuclear plants. As several user tasks with varied deadlines have to be executed in a certain interval of time, FPGA-based real-time task schedules are generated. Ensuring security for such FPGA-based real-time task schedules is of key importance. But like ASICs, real-time task operations in FPGAs can also be affected due to hardware attacks. Hence, security for real-time FPGA-based task operations is of utmost importance.

Hardware security is an arena that has been extensively researched for about two decades. This field gained significant momentum from 2005 when the US Government of Defence recognized hardware trojans as a significant threat to mission-critical applications. This arena was not only confined to research but has also been incorporated into various undergraduate and graduate courses. Moreover, with the cases of Meltdown and Spectre vulnerabilities in various microprocessors that are available in the electronics market, professionals who are associated with design and development of computing systems and their associated security have taken a keen interest in this arena of hardware security. However, the study on how hardware attacks may affect real-time task schedules is yet in its infancy. Not only detecting such vulnerability but also knowing the mechanisms of how such real-time task schedules can be protected from hardware attacks at runtime is of importance.

All species try to survive in nature and in this process, they exhibit a certain degree of self-awareness, based on which they counteract the various dangers and threats. Self-awareness (with respect to security) can be described as the ability of an individual to recognize itself in a complex environment, monitor or perceive the changes ongoing in the surrounding environment, decipher its course of action and finally, take an appropriate action to secure itself from the threat and survive in the environment. Based on this observe-decide-act mechanism, several on-chip self-aware security techniques were proposed in past works. However, the arena of self-aware security strategies that can secure real-time task schedules from hardware attacks is yet in its infancy and needs extensive study. Thus, the study of hardware security is incomplete, unless the effects of hardware attacks on real-time task schedules and its related self-aware security strategies are studied.

This book provides an extensive study on the latest FPGA-based scheduling strategies and how these can be affected due to hardware attacks. Along with these, self-aware security mechanisms for counteracting such threats at runtime are also presented in this book. This book will add on to the knowledge that are available in the existing books on real-time scheduling on reconfigurable hardware platforms and hardware security. This book will be useful to readers at all levels, be it students, research scholars, academicians or industry professionals, as it will serve to fill up the gaps in the existing knowledge on real-time scheduling on reconfigurable platforms and hardware security.

The book has three parts and is organized in the following manner:

Part I provides an introduction to real-time systems and discusses how they can be classified as hard and soft real-time systems,-based on different aspects and features. In addition to this, a detailed discussion is made on reconfigurable hardware or field-programmable gate arrays (FPGAs). Discussions are done from its evolution and conceptual background to its architecture and technology. Before conclusion, related discussions are made on fully and partially modes of reconfiguration of the FPGA fabric and spatio-temporal scheduling of hardware tasks on the FPGA platform.

Part II deals with scheduling of real-time tasks on FPGAs and comprises two chapters.

Chapter 2 gives a survey of real-time and rate-based resource allocation emphasizes its need and area of applicability. Different proportional fair algorithms have been discussed here. Finally, the chapter presents an overview of a few emerging scheduling considerations like energy awareness, imprecise computation, etc.

Chapter 3 discusses two novel real-time schedulers for scheduling periodic hard real-time task sets on fully and runtime partially reconfigurable FPGAs. A strategy along with associated theoretical analyses have been presented to show how an efficient scheduling strategy can be devised for FPGAs while correctly taking into account their inherent architectural constraints. Experiments conducted reveal that the proposed algorithms are able to achieve high resource utilization with low task rejection rates over a variety of simulation scenarios.

Part III is associated with security and comprises four chapters.

Chapter 4 is an introduction to hardware security, with special focus on reconfigurable hardware or FPGA-based platforms. It discusses hardware threats and the difference between hardware trust and hardware security. The lifecycle of an FPGA-based system is discussed, along with potential attack scenarios and existing security techniques.

Chapter 5 considers passive threats. In this chapter, it is analysed how confidentiality attacks may take place for fully and partially reconfigurable FPGA-based platforms that execute real-time task schedules. Development and working of self-aware agents are discussed that has the potential of bypassing such threats and facilitating self-aware security.

Chapter 6 focuses on active threats, i.e., integrity and availability attacks. Threat analysis is performed for single and multi-FPGA-based platforms associated with execution of real-time task schedules, that may operate in fully and partially reconfigurable mode. Self-aware agents that can detect such threats and mitigate them at runtime is discussed.

Chapter 7 considers power-draining attacks for embedded CPSs. Handling such attacks via a self-aware strategy is discussed in this chapter. Special focus is made on the handling of periodic and non-periodic tasks to ensure successful operations till the lifetime of the system.

We expect that the readers will enjoy reading the contents of the book and greatly benefit from them. We believe that the book contents that feature present age research will be highly relevant in the years to come and serve as a reference material for cutting-edge future research activities for academia and industries.

Kolkata, India   Krishnendu Guha
Colchester, UK   Sangeet Saha
Kolkata, India   Amlan Chakrabarti

Acknowledgements

The present book is the result of a long and arduous journey associated with several ups and downs. From planning and preparing the contents to finally scripting the unconventional research into an easy-to-read and understandable format is quite rewarding, from each of the author's viewpoints.

Most of the contents of this book are associated with topics that are results of fruitful Industry collaborations and successfully completed Government projects. Among all others, we wish to particularly mention the funding and support provided from the following:

Government

(i) Department of Science and Technology, Government of India

  - INSPIRE Program

(ii) Ministry of Human Resource Development (MHRD), Government of India

  - Special Manpower Development Program- Chip to System Design (SMDP-C2SD)

(iii) All India Council for Technical Education (AICTE), Government of India

  - Technical Education Quality Improvement Program, Phase III (TEQIP III)

Industry

(i) Intel India
(ii) Tata Consultancy Services (TCS)

First and foremost, the authors wish to thank all the past and present members of the Embedded and IoT Laboratory of A. K. Choudhury School of Information Technology, University of Calcutta, and Embedded and Intelligent System Laboratory, University of Essex, where they carried out their research work. Peer discussion, with valuable feedback from other lab members, is a key ingredient of the present book. Moreover, the memories shared during this time is always to be cherished by the authors.

The authors would like to especially thank Dr. Arnab Sarkar of IIT Kharagpur and Dr. Debasri Saha of University of Calcutta, Professor Klaus McDonald-Maier, University of Essex, for providing valuable suggestions related to the contents of this book. In addition to this, the authors would also like to thank Dr. Krishna Paul and Mr. Biswadeep Chatterjee of Intel India for providing significant insights and shaping up the book with relative contents that are important from the industry viewpoint and perspective.

The authors are grateful and would like to thank the Springer Nature editors and publishing team, in particular to Prasanna Kumar Narayanasamy and Annelies Kersbergen, for their patience, continuous support and guidance through the whole process.

Last but not the least, the authors would like to thank their parents and their family members for their constant support and encouragement.

Once again thanks to you all.

Krishnendu Guha
Sangeet Saha
Amlan Chakrabarti

Contents

About the Authors

Dr. Krishnendu Guha is presently an Assistant Professor (On Contract) at the National Institute of Technology (NIT), Jamshedpur, India. Prior to this, he was a Visiting Scientist in the Indian Statistical Institute (ISI), Kolkata, India, from December 2020 to February 2021. He was also an Intel India Research Fellow from December 2019 to December 2020. He has completed his Ph.D. from the University of Calcutta. In his Ph.D. tenure, he received the prestigious INSPIRE Fellowship Award from the Department of Science and Technology, Government of India and the Intel India Final Year Ph.D. Fellowship Award from Intel Corporations, India. He completed his M.Tech. from the University of Calcutta, where he was the recipient of the University Gold Medal for securing the First Class First Rank. His present research arena encompasses Real-Time Systems, Mixed Critical Systems, Embedded Security, with a flavour of Artificial Intelligence and Nature-Inspired Strategies.

Dr. Sangeet Saha received his Ph.D. degree in Information Technology from the University of Calcutta, India, in 2018. He received the TCS Industry Fellowship Award during his Ph.D. After submitting his Ph.D. thesis in 2017, he worked as a visiting scientist at the Indian Statistical Institute (ISI) Kolkata, India. Since May 1, 2018, he is a Senior Research Officer in EPSRC National Centre for Nuclear Robotics, based in the EIS Lab, School of Computer Science and Electronic Engineering at the University of Essex, UK. Primarily, his research expertise is in embedded systems, with specific interests that include Real-Time Scheduling, Scheduling for Reconfigurable Computers, Fault-Tolerance and Approximation-based Real-Time Computing.

Prof. (Dr.) Amlan Chakrabarti is presently Professor and Director of A. K. Choudhury School of Information Technology (AKCSIT), University of Calcutta. Prior to this, he completed his postdoctoral research at Princeton University after completing his Ph.D. from the University of Calcutta in association with ISI, Kolkata. He has been associated with research projects funded by government agencies and industries related to Reconfigurable Architecture, VLSI Design, Security for Cyber-physical Systems, Internet of Things, Machine Learning, Computer Vision and Quantum Computing. He is the Series Editor of Springer Transactions on Computer Systems and Networks and Associate Editor of Elsevier Journal of Computers and Electrical Engineering. His present research interests include Reconfigurable Computing, Embedded Systems Design, VLSI Design, Quantum Computing and Computer Vision. He is a Distinguished Visitor of IEEE Computer Society and Distinguished Speaker of ACM (2017–2020).

Part I
Introduction

Chapter 1
Introduction

1.1 Introduction

The growth in complexity and diversity as well as the end of the free ride in scaling up clock frequencies to achieve performance gains has led to a paradigm shift from single core embedded system designs to alternative architectures that allow better power and thermal efficiency along with increased hardware utilization (through the concurrent execution of multiple parallel threads) [KWM16]. Although homogeneous multi-core designs have received the most attention among these alternative architectures, continuing demands for higher performance, lower power budgets and fault-tolerance has inspired more complex heterogeneous system on chips (SoCs) oriented designs which often co-execute a mix of software and hardware components [SSS+14]. Present day android smart phones are an interesting example of such complex SoC based designs. It contains multiple general purpose processor cores (to execute the android operating system together with the plethora of software applications that come along with it), specialized digital signal processing (DSP) blocks, analog components such as radio frequency (RF) circuits (to communicate with the external world), a variety of micro electro mechanical systems (MEMS) (to sense its ambient world), dedicated digital hardware units (to implement performance critical functionalities such as video decoders, power management units) etc.

To further address the demands for lower costs, shorter times to market as well as flexible and adaptive application-specific instantiation, these hardware components are often implemented on re-configurable fabrics (FPGAs) integrated with the general-purpose processing elements to execute the software components. Tasks as software applications executing on general purpose processor exhibit high flexibility, but often at the cost of poor or very low performance. On the other hand, although tasks implemented as dedicated hardware circuits, i.e. application specific integrated circuits (ASICs), have high performance, they exhibit low flexibility and higher cost. Thus, re-configurable systems are emerging as a promising solution to amalgamate

K. Guha et al., *Self Aware Security for Real Time Task Schedules in Reconfigurable Hardware Platforms*, https://doi.org/10.1007/978-3-030-79701-0_1

the performance and flexibility where software tasks can run on a processor along hardware tasks that run on an FPGA. FPGAs empower systems by allowing different hardware-based application instances to be multiplexed on the same platform at different times.

Thus today, FPGAs are a bright prospect as platforms for executing even performance critical task sets, while allowing high resource utilization. However, very little research had been conducted towards effective spatio-temporal scheduling that achieves high resource utilization. A significant class of embedded system today are "real-time" in nature.

A real-time system has two notion of correctness, logical and temporal. Thus, a real-time system must not only be functionally correct, it must also produce result before a pre-specified time called deadline. An anti-lock braking system in a car is an example of typical real-time systems. In this system, in order to prevent the wheel from locking, the brakes need to be released within 200 ms.

1.2 Real-Time Systems

In order to describe "real-time" system, we need to emphasise on two important words, "real" and "time". "Time" denotes that the system's correctness will depend upon result generation time. "Real" denotes that the system must react towards some external events during their occurrence. Hence, all together we can say that a real-time system is characterized by it's necessity to satisfy two notions of correctness, *functional* and *temporal*. Therefore, such systems must not only produce correct results, i.e. functional correctness, but also the results should be produced before a stipulated time bound called *deadline* or ensure temporal correctness.

Real-time systems ranges over various domains, which includes systems for industrial control, automation, aviation, multimedia, consumer electronics, telecommunications, etc. A typical example of real-time system is provided by a temperature controller in a chemical plant that is required to switch off the heater within 30 milliseconds when the temperature reaches 250°, to avoid an explosion.

A typical real-time control system is shown in Fig. 1.1. The underlying control systems receives inputs from the external environment through various sensors and reacts to environment through the actuators. A fire alarm of a hotel room is an example of such a real-time system. The fire alarm has "smoke detection sensors" which always make a surveillance after small periodic intervals in the room (which in the present case is the environment), to detect whether there is any trace of smoke in the room. As soon as the sensors detect smoke, the control system sends command to the actuators that rings the alarm. It should be noted that all these events have to be completed within a timeline (which in the context of a real-time system is often denoted as *deadline*). A catastrophic consequence may occur if the control system operates late, after detecting the smoke.

Now, a question may arise in the reader's mind that "*Should all real-time systems act fast?*". The answer is: The word "fast" is relative. Rather than fast, it is preferable

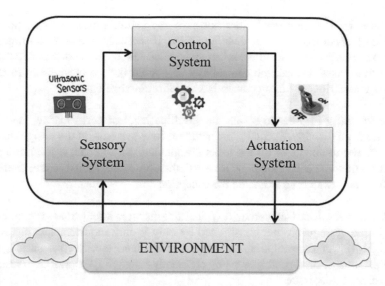

Fig. 1.1 A typical real-time control system

that real-time systems must act predictably. In order to achieve this predictability, from designing to testing phase of real-time systems, the new methodologies should be employed and investigated.

1.2.1 Hard Versus Soft Real-Time

The hardness of a real-time system is determined by the criticality of missed deadlines. Missing a deadline in a hard real-time system may lead to catastrophic consequences. Controllers for airplanes, monitoring systems for nuclear power plant, robot navigation systems are typical examples of hard real-time systems.

On the other hand, a soft real-time system is less restrictive; it can tolerate deadline misses at the cost of the quality of results, as long as they remain within certain temporal limits beyond which the system becomes useless. Obviously, Quality of Service (QoS) degrades as delay in response increases beyond deadline. Examples of such systems include streaming video, voice over IP, interactive gaming etc.

1.2.2 Important Features of Real-Time Systems

Irrespective of whether a system is soft real-time or hard real-time, the real-time system should exhibit the following properties:

- *Timeliness* A correct result does not have any value, until it is being produced within a specified time, i.e. deadline. Now, if the tasks are soft real-time, then up to a certain extent, deadline miss can be tolerated, but it is not possible for hard real-time tasks. Hence, the control system should have a mechanism to handle tasks with stringent timing constraints and different criticality.

- *Predictability* In order to obtain the specified level of performance, the system must act predictably. Specially for hard real-time systems, at the design time, the timing requirements of all the tasks are guaranteed. In case, if the timing constraints of some tasks cannot be satisfied, then the system must ensure alternative actions at design time to handle the exception.

- *Efficiency* As discussed earlier, a typical fire-alarm is a real-time system. Hence, real-time systems are not constrained by only timing requirements, but also by weight, power consumption, memory management, etc. So, for such systems achieving a desired performance by efficiently managing the available resources is a challenging issue.

- *Robustness* Based on the situation, a real-time system may suffer extremely high load, if some external events happen. Hence, its performance should not deviate when the system load changes. A proper real-time system must behave robustly if there is certain load variations.

- *Fault tolerance* Any typical electronic system may suffer faults, so does the real-time systems. However, a real-time system must work properly in spite of having software or hardware faults. Thus, each real-time system is backed up with proper fault tolerant techniques so that the tasks do not miss the deadlines.

- *Maintainability* Based on external environment, the specification of a real-time system can be changed. Hence, such systems should be designed in such a way (preferably in modular structure) so that any modifications can be performed easily.

1.2.3 Real-Time Tasks and Its Classifications

A task is composed of a set of instructions to be executed on a processor. Typically, a task executes repeatedly and each such execution instance is referred to as a job. Important parameters which characterize a real-time task are:

- a_i: Arrival time of a task T_i
- e_i: Execution requirement or time required to complete execution of a job of T_i
- R_i: Release time of job or the time at which the job of T_i becomes ready to execute
- D_i: Absolute deadline of a job or the time instant by which the job of T_i must complete execution
- d_i: Relative deadline of a job of T_i, i.e. $D_i - R_i$
- r_i: Response time or the time at which a job of T_i finishes execution relative to its release time R_i.

Periodic, Aperiodic and Sporadic tasks: Tasks are classified as periodic, aperiodic or sporadic based on their arrival pattern.

Aperiodic tasks does not follow any restriction on the arrival of their jobs. Two consecutive jobs of a sporadic task must be separated by a minimum inter-arrival time, while in case of periodic tasks this inter-arrival time is fixed.

Precedence Constrained Tasks: Many embedded real-time applications are conveniently modeled as precedence constrained task graphs. A precedence constrained task graph $G = (T, E)$ is composed of a set T of task nodes and a set E of edges between task nodes. Each task node $T_i \in T$ represents a distinct functionality of an application. Each node $e_{ij} \in E$ denote precedence constraints between a distinct pair of task nodes T_i, T_j. As an example, let us consider the Adaptive cruise control (ACC) system in a modern car whose objective is to maintain a safe distance with other automobiles in the vicinity. Figure 1.2 shows the model of ACC which shows the inter dependency among its various components. ACC first determines the current distance and speed of an automobile in the neighborhood along with the speed at which it itself is running. Based on these parameters ACC determines the desired speed followed by the required braking force and throttle position (relative to current throttle position). Finally, it generates actuation outputs to actual physical controls over throttles and brakes. Of course, this entire set of activities must be performed within stringent timing constraints which is modeled through an overall deadline for the completion of all activities of the application graph.

Preemptive and Non-Preemptive tasks: A non-preemptive task is that which once started, cannot be interrupted before the completion of its execution. Preemptive tasks on the other hand, may be interrupted before completion to possibly allow the execution of other tasks.

Extensive research has been conducted towards scheduling in both hard and soft real-time systems. While hard real-time scheduling strategies are based on exact guarantees towards meeting deadline, soft real-time scheduling mechanisms typically need to provide statistical assurances on deadlines. In the next few subsection, we take a deeper look at scheduling mechanisms in real-time systems.

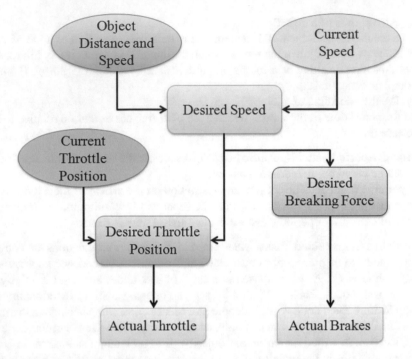

Fig. 1.2 ACC block diagram

1.3 Field Programmable Gate Arrays (FPGAs)—Its Evolution and Conceptual Background

1.3.1 Introduction to FPGAs

FPGA is an electronic device which consists of a matrix of re-configurable logic circuitry, typically referred to as *configurable logic blocks* (CLBs), surrounded by a periphery of I/O blocks, as shown in Fig. 1.3. When a FPGA is configured, the internal circuitry forms hardware implementation of the desired application. Unlike processors, FPGAs use dedicated hardware (using CLBs as building blocks) for processing logic. Hence, FPGAs are truly parallel in nature and do not have to compete for the same resources when processing different operations of an application, as is the case with a software implementation using processors. As a result, performance of different components of the application become mutually independent; additional processing incorporated in one component do not affect the performance of other components.

However, unlike single-purpose hardware designs (also referred to as, *Application Specific Integrated Circuits or ASICs*) which has dedicated and fixed hardware functionalities, FPGAs can literally re-wire their internal circuitry to allow post-

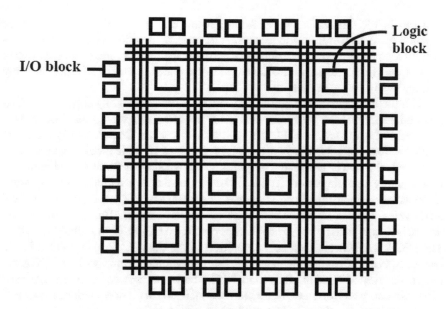

Fig. 1.3 Simple FPGA internal architecture [BEGGK07]

fabrication hardware programmability and thus, these devices are the enabling technology for re-configurable computing. By incorporating programmability, FPGAs are able to combine the flexibility of software based implementations along with performance efficiencies that are often close to that of single purpose hardwares or ASICs.

A side effect of the power of reconfigurability is the ability to correct/modify design errors even after the implementation phase. This is unlike ASICs where it is almost impossible to rectify design errors post fabrication. This restriction increases the design, verification and testing overheads of ASICs by multiple folds, in terms of both *cost* and *time-to-market*.

With the objective of alleviating such overheads, FPGA based implementations are often preferred over ASICs in situations where the intended device is not targeted for huge mass production. Due to the same reason, FPGAs find wide popularity as prototyping platforms before the mass manufacturing phase. However, power and area related overheads for FPGAs are typically higher than those for ASICs and this limits the design size.

1.3.2 FPGA Technology: Evolution of Its Conceptual Path

The first user-programmable chip capable of implementing logic circuits was the Programmable Read-Only Memory (PROM). Later, PLA, programmable logic devices replaced the PROMs. Basically a PLA implements Boolean functions in SOP (sum of product) form. However, due to two levels of programmability PLA introduced significant amount of propagation delays. To counter this shortcomings, Programmable Array Logic (PAL) devices were developed. In contrast to PLA, PALs provide only a single level of programmability.

With the advancement of technology, PALs started containing flip-flops so that sequential circuits can be realized. These are often referred to as Simple Programmable Logic Devices (SPLDs). As the time passed, Complex Programmable Logic Devices (CPLDs) were evolved. CPLD is a set of logic blocks distributed in macrocells that can be connected programmatically using an interconnection matrix based mainly on multiplexors. The logic blocks available on a CPLD are very similar to those available on a Programmable Logic Device (PLD), a matrix of OR and AND gates that can be connected to obtain digital circuits. The logic blocks are themselves comprised of macrocells. Macrocells on CPLDs also include registers. The FPGA was conceived of as an evolution of the Complex Programmable Logic Device (CPLD). The main characteristics of the evolution from CPLD to FPGA are the greater flexibility of the interconnectivity between logic cells, the inclusion of SRAM memory bits (discussed later), and the inclusion of other embedded resources, such as multipliers, adders, clock multipliers, and so on.

1.3.3 FPGA Architectures

A basic FPGA consists of three major elements as shown in Fig. 1.4

1. Combinational logic blocks
2. Programmable interconnects
3. Programmable Input/Output pins.

Combinational logic blocks are configurable blocks used to implement custom logic functions. They are denoted as logic elements (LEs) or configurable combinational logic blocks (CLBs). These blocks are generally arranged as a two-dimensional structure and are surrounded by programmable interconnects. The design to be implemented is divided into small modules, each fitting in a logic block. Several blocks are then interconnected using programmable interconnects in order to implement the whole circuit (design). As the FPGA is re-programmable, the implemented function could be changed by updating the content of the configuration memories.

1.3.4 Heterogeneous FPGAs

Today, FPGAs are increasingly used in many computationally intensive applications with stringent performance requirements. In order to satisfy performance demands,

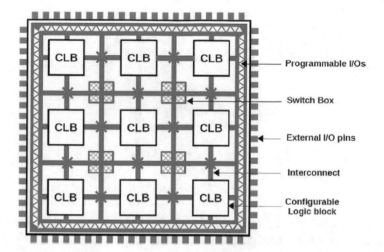

Fig. 1.4 Detailed FPGA internal architecture [BEGGK07]

Fig. 1.5 Heterogeneous
FPGAs: a conceptual block
diagram [BEGGK07]

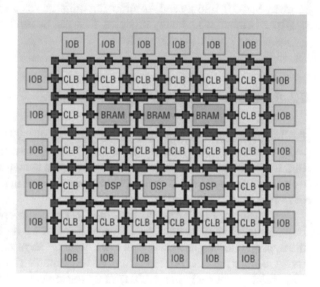

these FPGAs often include specialized embedded Hard Blocks (HBs) such as memory blocks and DSP units within the uniform matrix of homogeneous CLBs, as shown in Fig. 1.5. Inclusion of these HBs enable present FPGA platforms to be more heterogeneous in nature. The most common embedded HBs are:

– Memory Blocks: In applications like image processing, huge amounts of intermediate data need to be frequently and temporarily stored during their processing. Use of embedded memory blocks in FPGAs have become crucial for the efficient implementation of these applications with reduced memory access delays. These

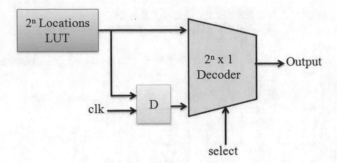

Fig. 1.6 Configurable logic block

memory blocks, often called Block RAMs (BRAMs) provide dedicated storage capacity upto about ≈ 1 MB in many modern FPGAs.

– Embedded DSP Blocks: Embedded DSP blocks in FPGAs commonly provide many MAC (Multiply and Accumulate) units. Associated with the aforementioned BRAM embedded memory modules, they can be used to easily implement digital processing functions such as filters.

1.3.5   Closer Look into CLBs

A CLB, which may be considered as the unit building blocks of an FPGA, is essentially composed of a Look Up Table (LUT), a register and a decoder as shown in Fig. 1.6. The register shown in the figure is used to synchronize the LUT output with a clock, if necessary. A LUT with 2^n locations along with a $2^n \times 1$ decoder is capable of implementing any *n-input* function.

For example, Fig. 1.7 shows how the LUT-decoder combination can be used to implement the boolean function $F = a \times b + \overline{a} \times c$ over inputs (a, b, c), by storing the appropriate output values in a LUT with 2^3 locations. It is clear that 2^{2^n} different *n-input* boolean functions may be implemented by programming and storing appropriate values within a LUT with 2^n locations and irrespective of the function implemented, the delay to produce an output remains same.

1.3.6   FPGA Design Flow

The FPGA design flow begins with the behavioral description of the intended application using a Hardware Description Language (HDL) like Very High Speed Integrated Circuit Hardware Description Language (VHDL)/Verilog or a schematic capture environment. This step of the design flow is interspersed with periods of func-

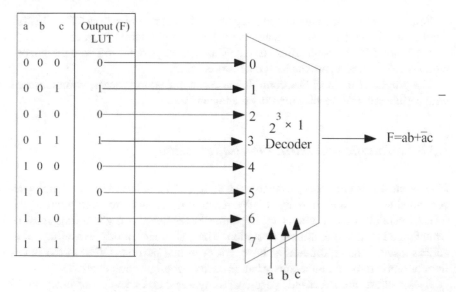

Fig. 1.7 Combination of a 2^3-location LUT with a $(2^3 \times 1)$ decoder to implement the function $F(a, b, c) = ab + \bar{a}c$, within a simple CLB

tional simulation to verify the correctness of the intended design. After this step, the designer can find that whether the logic is functionally correct or not before proceeding to the next stage of development.

At the next step, the design gets synthesized into an intermediate representation called *netlist*. The generated netlist is then passed through a translation process called *place & route*. This step maps the netlist onto actual macrocells, interconnections, and input/output pins. As the output of place & route, *bitstream* is generated.

Bitstream is the configuration data that needs to be loaded in the configuration memory of the FPGA to implement the desired design.

The generated bitstream may be downloaded into the FPGA through configuration interfaces like joint test action group (JTAG) port, SelectMap or Slave Serial ports and internal configuration access port (ICAP). These ports enable numerous re-configuration techniques including compressed, encrypted and partial bitstream download and readback.

1.3.7 Processors Within Re-configurable Targets

A processor core in an FPGA not only improves the flexibility of the system but also provides a support for software execution of an application. In Xilinx FPGAs, we can find two types of processors: soft-core and hard-core.

Soft-core processors, are implemented using FPGA resources (like CLBs, LUTs). On the other hand, hard-core processors are hard-wired on the FPGA chip.

For Xilinx FPGAs, MicroBlaze or PicoBlaze are the popular soft core processors, while ARM, PowerPC are the hard core processors.

The processor in an FPGA controls the reconfiguration process, communicates with peripherals and handles memory management.

1.3.8 Dynamic and Partial Reconfiguration

FPGAs are most commonly categorized as fully re-configurable or partially re-configurable platforms. Initially, FPGAs came only as fully re-configurable platforms where all logic resources on the entire FPGA floor at a given time must be reconfigured as a single atomic operation. Thus, all concurrently executing applications instantiated on different sub-regions or virtual portions (VPs) of the FPGA floor or fabric must simultaneously halt at each re-configuration event.

Such a constraint restricted space-shared dynamic co-execution of independent applications running in different VPs within the area of the floor. However, such a restriction was removed with the advent of commercial platforms that also feature dynamic partial reconfiguration enabling a hardware instantiation running in a VP, to be reconfigured while allowing instantiations in other sub-regions to continue execution uninterruptedly.

Xilinx FPGAs usually adopt the *module-based style* for partial reconfiguration. A *module* in a module-based reconfiguration is determined by a separate partial bitstream, also called a "task". Such a module in a particular FPGA sub-region or VP may be reconfigured without interrupting the execution of other modules, by downloading the module's bitstream through the ICAP into the configuration memory of the sub-region or VP (Fig. 1.8).

Fig. 1.8 Programmable logic design process

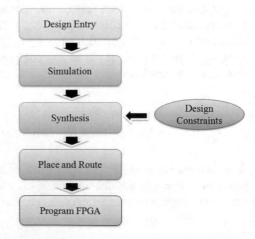

1.3.9   Real-Time Hardware Tasks

A software task is defined as a set of instructions (a piece of code) and data (i.e. handled by those instructions) which is executed on a processor. Important parameters that characterize a real-time software task include execution time, deadline, response time, period, latency etc.

Similarly, with respect to application execution on FPGAs, we bring in the concept of a hardware task. FPGAs contain a re-configurable resource which basically consists of a 2-dimensional array of $W \times H$ CLBs. A hardware task T_i in such a system is a relocatable digital circuit, typically rectangular in shape, which may be configured to be executed anywhere consuming a sub-region $w_i \times h_i$ on the floor (having total area $W \times H$) of the re-configurable device. Such a sub-region is also termed as a virtual portion (VP) of the FPGA. In addition to these geometric features, all parameters which characterize a real-time software task, also apply to real-time hardware tasks.

1.3.10   Spatio-temporal Management of Hardware Tasks

Parallel execution of hardware tasks can be achieved by placing them simultaneously on the FPGA floor provided no task (region) overlaps with other tasks or with the boundary of FPGA. After placing a set of tasks, if the remaining vacant region is not enough to place any other tasks then the place can be considered to be wasted, due to *fragmentation*. Hence, *fragmentation* minimisation remains one of the principal goals of any placement strategy. A *Placer* assumes the responsibility of placing a task onto the FPGA floor after verifying that sufficient residual spatial resources are available to accommodate the task. If a feasible sub-region to place the task is found, then a *Loader* downloads the task's bitstream through ICAP onto the configuration memory of the FPGA.

It can be observed that all tasks can not be accommodated simultaneously. It is then necessary to additionally multiplex tasks over time on the available spatial resources. *Scheduler's* responsibility in such a system therefore, is to decide both *where* and *when* to execute tasks in the system.

Thus, the job of a real-time scheduler for FPGAs is to efficiently manage both space and time while appropriately accounting for reconfiguration related overheads, such that all deadlines are met. Hence, a spatio-temporal scheduling algorithm simultaneously conduct both temporal allocation along with placement. To specific issues of importance for a dynamic scheduler may be pointed out here.

– Which tasks should be selected for execution at a scheduling point?
– What are the best feasible regions available to place those tasks?

1.3.11   Various Task Placement Strategies for FPGAs

The generic problem task placement deals with two objectives i.e.
(i) Minimize the area consumed by placed tasks so that new tasks can be placed.
(ii) Within a FPGA floor maximize the number of placed tasks.

However, tackling both the problems appear as NP-complete in the strong sense [CCP06]. Therefore, researchers are inclined towards developing various heuristic placement approaches [TSMM04, SWPT03, Mar14]. Majority of such heuristics [HSH09] has attempted to place tasks by following different restricted area models. While placing a task, it is already discussed that due to the limited resources of the FPGA, all tasks can not be accommodated simultaneously. Hence, a task can only be placed if sufficient vacant space is available. In order to handle the placement of task on FPGA there are two main components i.e.

– **Placer**: Scheduler employs scheduling algorithm and finds out the task that needs to be executed. Once this information is available, placer will be responsible to find out the feasible position on the FPGA. If there exists a feasible position, the placer will send the command "loader" which will then load the new bitstream through ICAP. Once, the tasks are loaded, the remaining free areas will be updated.

– **Free Space Manager**: A free space manager keep track of the unoccupied area. It keeps communication with the Placer so that the free resources can be updated whenever a task is placed or removed from the FPGA.

Task Placement for 1D Area Model In this area model, re-configurable width W of the floor of the FPGA is divided into a constant number of equal sized columns (termed as *tile*), as shown in Fig. 1.9. Any upcoming task will be placed in any of these tiles. It is worth mentioning that FPGAs does not have any such physical regions so these are often called as virtual partition or VP. We have used VP and tiles interchangeably through out this book.

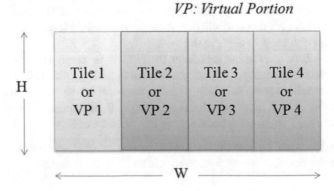

Fig. 1.9 1D Area Model

The authors in [SWP03] introduce the horizon and stuffing techniques for 1D model. These techniques always place the arriving tasks from the leftmost column. However, as the columns have equal width, so if the width of the placed task is smaller than the tile width then it will cause severe fragmentation.

In order to handle the fragmentation problem, Chen et al. [CH05] propose a task placement method called "classified stuffing". The have coined a new parameter called SUR and the arriving tasks are placed on the basis of SUR. SUR is defined as the ratio between the area requirement and the execution requirement of a task i.e. it is basically a ratio between area and time. For a task, if $SUR > 1$, the the task is placed in the leftmost available columns of the FPGA and if $SUR < 1$ then it is placed in the rightmost available columns.

Hubner et al. [HSB06] proposed the idea of vertical placing of tasks within a tile i.e. place different tasks on top of each other within a tile, until and unless the sum of the heights of these tasks do not exceed the height of the slot. The authors succeeded in some extent to reduce the fragmentation.

Task Placement for 2D Slotted Area Model An improvement over the 1D slotted area approach is the 2D partitioned area model where both the width W and height H of the floor of an FPGA is partitioned into equal intervals to obtain a fixed number of equal sized tiles/VPs [YSL12], refer Fig. 1.10. Tiles are place-holders for hardware tasks and each tile can accommodate no more than one hardware task at a time.

As slotted 2D area model uses equal sized tiles hence, it may suffer from fragmentation if the size of the task is less than the size of the tile. Hence, one improvement over slotted area model becomes flexible 2D area model which allows a task to be placed onto any arbitrary region of the FPGA. However, though this placement strategy provides compact placement, but it involves much higher computational complexity. In general, relaxations on the 2D slotted area model (in order to provide higher resource utilization), lead to two important drawbacks:

– Even though they improve average performance for a set of best-effort tasks, they
  often degrade scheduling predictability which is of utmost importance, especially

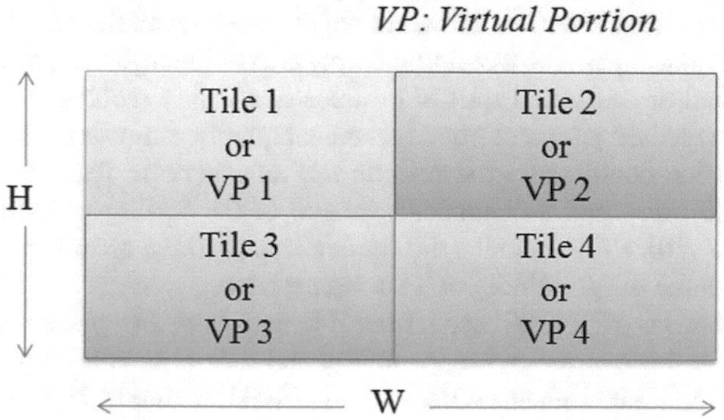

Fig. 1.10 2D slotted area model

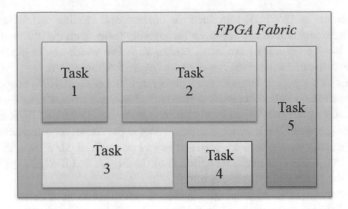

Fig. 1.11   2D flexible area model

in real-time systems. This is because, it becomes more complex in this case to deterministically account for the spatio-temporal capacity available in the worst-case, within a given time interval in future.
– These models tend to incur much higher overheads at each spatial scheduling point, making the complexity of the overall spatio-temporal scheduling problem is very expensive towards online application.

Task Placement for Flexible 2D Area Model This is the most flexible model that allows allocation of tasks anywhere on the FPGA as shown in Fig. 1.11. By following this model, tasks can be placed in compact manner which in turn increases resource utilisation. However, the high flexibility of this model makes scheduling and placement more difficult. But this The 2D area model also suffers from fragmentation. If tasks are placed on any arbitrary positions then the area in between two tasks remain unutilised.

In [BKS00], Bazargan et al. proposed KAMER (Keeping All MERs), a mechanism which proceeds by maintaining a list of Maximal Empty Rectangles (MERs) that cannot be covered fully by a set of other empty rectangles. Whenever a task arrives, the algorithm places it in the bottom-left corner of the largest available MER in a worst-fit manner. After placement, the remaining empty area of the region is partitioned using either a vertical or horizontal split to produce two empty rectangular sub-regions. Although this method produces good placement quality, a drawback is that a wrong splitting decision could cause the rejection of an otherwise feasible task. Walder et al. [WSP03] proposed an enhanced version of the Bazargan partitioner which postpones the vertical/horizontal splitting decision until the arrival of a new task in order to overcome the possibility of a wrong decision.

The authors in [EFZK08] used First Fit and Best Fit placement strategies and claimed to achieve lower fragmentation and task rejection rates compared to KAMER [BKS00] and Enhanced Bazargon [WSP03] methods. However, this work maintains the occupancy status of each CLB of an FPGA in a 1D array and therefore, is susceptible to high computational overheads. A few other placement approaches

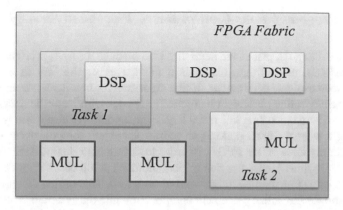

Fig. 1.12 2D Heterogeneous FPGAs floor area

aimed towards the efficient management of empty regions include algorithms by
Handa et al. [HV04] (which uses a staircase data structure to discover empty spaces)
and Tabero et al. [TSMM04] (which employs a Vertex List Set (VLS), where a given
free space fragment is represented by a list of vertices). Ahmadinia et al. [ABBT04]
proposed management of occupied area instead of free area as they observed that
records for occupied spaces grow at a much slower rate than those for free spaces,
making data management for the accommodation of new tasks simpler.

Task Placement for 2D Heterogeneous FPGAs Along with CLBs, FPGAs also
contain other embedded components like BRAM blocks, multipliers and DSPs. Now,
if a task requires such components for its execution then it cannot be placed anywhere
inside the FPGA. If a task needs a DSP block for its computation then while placing
the task it has to be ensured that the DSP lies within the task region, as shown in
Fig. 1.12. Hence, this heterogeneity imposes additional placement constraints for the
task.

In [KPK05], authors proposed an algorithm to deal with heterogeneous hardware
constraints while placing a task. Feasible placement positions of the given hardware
tasks are determined offline. At run-time, a task is placed at the first available free
position. A generalization over this problem model is proposed in [ECPS10], where
a task can have multiple instances with the scheduling objective being minimization
of task rejection ratio. Authors in [SER13] used a 2D slotted area model and classified tasks according to their resource requirements. The tasks with the maximum
required resources in a given class represents a slot in the FPGA. Hence this slot can
accommodate any task belonging to the class.

In [KCH14], the authors assume that the BRAMs, DSP are symmterically spaced
i.e. a BRAM blocked is placed after a certain distance from another BRAM and
similarly for DSP. If a new task arrive with BRAM requirements, then the proposed
algorithm will scan the unoccupied BRAMs and shall place the task.

1.3.12   Fragmentations Control Based Placement Strategies

The main challenge of placement heuristics is to manage the free space and schedule arriving tasks so that the fragmentation is avoided. This is quite difficult to achieve in an online scenario. As stated earlier, offline algorithms or heuristics are very efficient but slow, while online placement heuristics are faster with less efficiency. In some systems with some recurring idle times (e.g. in an automotive electronic system during the night when the car is parked), an offline placement component can optimally relocate the modules on the FPGA, in order to defragment the available free space during the idle time. Unused or non-critical modules could be interrupted and relocated in order to free the maximum contiguous space. This offline defragmentation eases the job for the online placement component while the system is in use. Hence, one can overcome fragmentation problems in some specific application by combining online placement with an offline defragmentation. Veen et al. [vdVFA+05] present such a strategy. Their online placer has an offline component called the defragmenter which performs an offline relocation of the currently placed tasks with the objective to maximize the areas of contiguous free resources. This defragmentation eases the online tasks placement process which is done at runtime. This approach is extremely useful for FPGA in which partial reconfiguration can only be performed columnwise (e.g. Xilinx Virtex II Pro FPGAs). Indeed, in such an FPGA, while performing a 2D placement, reconfiguring (or placing) a module on the FPGA affects all the modules interfering columnwise. If defragmented online, one has to reconfigure all the interfering modules, which is not desirable at runtime.

In order to limit the fragmentation of the re-configurable array, Ahmadinia et al. [ABBT04] proposed an algorithm that uses two horizontal lines to manage free spaces on the array. Instead of maintaining a list of empty rectangles, they managed two horizontal lines, one above and one under the already running tasks. The overall idea behind this approach is to put beside each other tasks with nearly the same finishing time. Hence, contiguous spaces are free at nearly the same time, reducing the fragmentation. The resulting empty space is then more likely to fit future and even larger tasks. Furthermore the principle is extended to a clustered approach, where the re-configurable array is split in a number of 1D clusters, each cluster accommodating tasks with similarities in terms of finishing time.

Koester et al. [KPK05] adopted a defragmentation-by-modules-relocation approach to deal with continuous fragmentation of the re-configurable array over time. As hardware tasks could be placed and removed at runtime, an increasing fragmentation of the FPGA prevents next tasks from being placed. Their solution is to relocate at runtime the currently placed tasks for being able to place the requested task. Their runtime defragmentation algorithm aims at implementing such an approach. Prototyped on a dynamically and partially re-configurable Xilinx Virtex-II FPGAs, their algorithm applied to the 1D placement shows some improvement of placement quality. For example, the total execution time of tasks set is reduced to 87.1% in the best case, compared to another scheduling algorithm without any runtime defragmentation. However, using this deframentation-by-relocation approach is worthy only if tasks reconfiguration time is negligible compared to tasks execution time. In addition, as noticed earlier, their approach takes into account the heterogeneity of the FPGAs by identifying feasible positions for tasks, according to their types of resources (logic cells or embedded memory).

Cui et al. [CDHG07] proposed an online task placement algorithm which aimed at minimizing fragmentation on partially re-configurable FPGAs. They introduced a 2D area fragmentation metric that takes into account the probability distribution of sizes (width and height) of future tasks arrival. Hence, in their assumption, dedicated embedded applications are targetted and consequently, task arrival patterns are predictable.

1.4   Conclusions

In this chapter, we have discussed that the real-time systems have a dual notion of correctness—logical and temporal. Ensuring temporal correctness i.e. meeting deadlines is ultimately a scheduling problem. CPUs or general purpose processors are the traditional choice for executing real-time tasks/applications. However, FPGAs are becoming a lucrative platforms for many of todays real-time systems. This chapter has explored many issues in the placement of real-time tasks on FPGA devices, both the homogeneous and heterogeneous FPGAs.

It has been observed that majority of the existing research on scheduling and placement of real-time tasks on FPGAs considered 1D or 2D homogeneous FPGA. The main objective for task placement remains to reduce the FPGA fragmentation, i.e. to place a large number of tasks in a compact way. Moreover, with the increasing complexity of tasks, tasks may also require BRAMs, DSPs, etc., and thus, the placement strategies should also deal with this heterogeneity during the task placement.

References

[ABBT04]   A. Ahmadinia, C. Bobda, M. Bednara, J. Teich, A new approach for on-line placement on reconfigurable devices, in *18th International Parallel and Distributed Processing Symposium, 2004. Proceedings.* (IEEE, 2004), p. 134

[BEGGK07] D. Buell, T. El-Ghazawi, K. Gaj, V. Kindratenko, High-performance reconfigurable computing. Comput.-IEEE Comput. Soc. **40**(3), 23 (2007)

[BKS00]   K. Bazargan, R. Kastner, M. Sarrafzadeh, Fast template placement for reconfigurable computing systems. IEEE Design Test Comput. **17**(1), 68–83 (2000)

[CCP06]   D. Chen, J. Cong, P. Pan, Fpga design automation: a survey. Foundat. Trends Electron. Design Autom. **1**(3), 139–169 (2006)

[CDHG07]   J. Cui, Q. Deng, X. He, Z. Gu, An efficient algorithm for online management of 2d area of partially reconfigurable fpgas, in *Proceedings of the Conference on Design, Automation and Test in Europe* (EDA Consortium, 2007), pp. 129–134

[CH05]   Y.-H. Chen, P.-A. Hsiung, Hardware task scheduling and placement in operating systems for dynamically reconfigurable soc. Embedded Ubiquit. Comput.-EUC **2005**, 489–498 (2005)

[ECPS10]   A. Eiche, D. Chillet, S. Pillement, O. Sentieys, Task placement for dynamic and partial reconfigurable architecture, in *(DASIP), 2010* (IEEE, 2010), pp. 228–234

[EFZK08]   M. Esmaeildoust, M. Fazlali, A. Zakerolhosseini, M. Karimi, Fragmentation aware placement algorithm for a reconfigurable system, in *Second International Conference on Electrical Engineering, 2008. ICEE 2008* (IEEE, 2008), pp. 1–5

[HSB06]   M. Hubner, C. Schuck, J. Becker, Elementary block based 2-dimensional dynamic and partial reconfiguration for virtex-ii fpgas, in *20th International Parallel and Distributed Processing Symposium, 2006. IPDPS 2006* (IEEE, 2006), p. 8

[HSH09]   P.-A. Hsiung, M.D. Santambrogio, C.-H. Huang, *Reconfigurable System Design and Verification*, 1st edn. (CRC Press Inc., Boca Raton, FL, USA, 2009)

[HV04]   M. Handa, R. Vemuri, An efficient algorithm for finding empty space for online fpga placement, in *Proceedings of the 41st annual Design Automation Conference* (ACM, 2004), pp. 960–965

[KCH14]   Q.-H. Khuat, D. Chillet, M. Hubner, Considering reconfiguration overhead in scheduling of dependent tasks on 2d reconfigurable fpga, in *2014 NASA/ESA Conference on Adaptive Hardware and Systems (AHS)* (IEEE, 2014), pp. 1–8

[KPK05]   M. Koester, M. Porrmann, H. Kalte, Task placement for heterogeneous reconfigurable architectures, in *2005 IEEE International Conference on Field-Prog. Tech., 2005. Proceedings* (2005), pp. 43–50

[KWM16]   D.B. Kirk, W. H. Wen-Mei, *Programming Massively Parallel Processors: A Hands-on Approach* (Morgan Kaufmann, 2016)

[Mar14]   T. Marconi, Online scheduling and placement of hardware tasks with multiple variants on dynamically reconfigurable field-programmable gate arrays. Comput. Elect. Eng. **40**(4), 1215–1237 (2014)

[SER13]   M. Sanchez-Elez, S. Roman, Reconfiguration strategies for online hardware multitasking in embedded systems. arXiv:1301.3281 (2013)

[SSS+14]   V. Sklyarov, I. Skliarova, J. Silva, A. Rjabov, A. Sudnitson, C. Cardoso, Hardware/software co-design for programmable systems-on-chip (2014)

[SWP03]   C. Steiger, H. Walder, M. Platzner, Heuristics for online scheduling real-time tasks to partially reconfigurable devices. Field Program. Logic Appl. 575–584 (2003)

[SWPT03]   C. Steiger, H. Walder, M. Platzner, L. Thiele, Online scheduling and placement of real-time tasks to partially reconfigurable devices, in *24th IEEE Real-Time Systems Symposium, 2003. RTSS 2003* (IEEE, 2003), pp. 224–225

[TSMM04]   J. Tabero, J. Septién, H. Mecha, D. Mozos, A low fragmentation heuristic for task placement in 2d rtr hw management, in *FPL* (Springer, 2004), pp. 241–250

[vdVFA+05] J. van der Veen, S.P. Fekete, A. Ahmadinia, C. Bobda, F. Hannig, J. Teich, Defragmenting the module layout of a partially reconfigurable device. arXiv:cs/0505005 (2005)

[WSP03]   H. Walder, C. Steiger, M. Platzner, Fast online task placement on fpgas: free space partitioning and 2d-hashing, in *International Parallel and Distributed Processing Symposium, 2003. Proceedings* (IEEE, 2003), p. 8

[YSL12]   L.T. Yang, E. Syukur, S.W. Loke, *Handbook on Mobile and Ubiquitous Computing: Status and Perspective* (CRC Press, 2012)

Part II
Scheduling

Chapter 2
Real-Time Scheduling: Background and Trends

2.1 Introduction

As we discussed in our previous chapter, predictability of a real-time system is an important feature. Several strategies and algorithms have been proposed to ensure the predictability. In order to understand the concept of predictability, we need to define the key concept of any real-time system i.e. "Task". We have mentioned the word "task" or specifically "real-time tasks" in the previous chapter. In this chapter we will have a detailed look. Task is synonymous with the word "thread" in operating system i.e. a sequential computation that is carried out by CPU.

In a real-time system, different task could have different criterion (i.e. hard deadline or soft deadline) hence, these tasks will be assigned to the CPU according to predefined policy which is called "*scheduling policy*". Now, scheduling algorithm will be responsible to determine the order in which tasks will be executed. When a scheduling algorithm allocates a task to a CPU then this operation is called as dispatching.

When a task is executing on a processor then it is called as running task and when a task is waiting for the processor then it is called ready task. All ready tasks are stored in a *queue* before they are allocated to processors. In a real-time system as we have discussed tasks might have different priorities and thus, a running task with lower priority can be interrupted by a higher priority task. Scheduling algorithm will attempt to allocate the higher priority task to the processor so that the higher priority task does not miss its deadline. This operation of suspending a running task and start execution of another tasks is called preemption. Figure 2.1 illustrates the concept stated above.

In a real-time system, preemption can be useful for three cases.

- Efficient exception handling: As we discussed in the previous chapter that real-time system should have exception handling mechanism. Hence, the tasks responsible for exception handling should be able to preempt existing tasks to issue exceptions in timely manner.

K. Guha et al., *Self Aware Security for Real Time Task Schedules in Reconfigurable Hardware Platforms*, https://doi.org/10.1007/978-3-030-79701-0_2

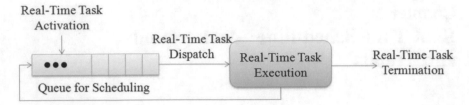

Fig. 2.1 Steps of task executions

- Criticality maintaining: As real-time system handles tasks with different levels of criticality hence, preemption allows executing the most critical tasks so that the deadline can be satisfied.
- Achieving high utilisation: Scheduling algorithms with preemption allows executing real-time tasks with higher CPU utilization.

Later in this chapter, we will observe that though preemption in case CPUs is easily achievable but in case of FPGAs, the preemption of a hardware task is a challenging issue.

2.2 A Background on Scheduling

As we discussed, scheduling may be considered as the problem of assigning a set of tasks to specific points on the time scale of an processor. Based on this basic definition, various paradigms of the problem have developed primarily based on the following criteria:

- The characteristics of tasks to be processed: Tasks may be static in nature or may be dynamic. In a static scenario, a certain number of tasks arrive simultaneously in a system that is idle and immediately available for execution. As it is assumed that no further tasks will arrive, scheduling decisions may be taken statically (at design time). However, in case of dynamic scenario, tasks may arrive and depart at anytime. Here, scheduling decisions must be taken dynamically at run time. Based on the situation, the tasks may or may not. Non-preemptive schedulers assumes that tasks will not be interrupted within its execution period. Hence, if the task has large execution times, it may take long time to response to external events. However, some tasks are by nature non-preemptive.
- Types of processing elements: If there is a single available CPUs or processing resource, the task set must be scheduled using a uni-resource scheduler, while presence of multiple resources requires a multi-resource scheduler. If all the resources are of the same type (homogeneous), then any task may be allocated to any resource. On the other hand, if this is not the case, if the processing resources have different characteristics (heterogeneous), then the tasks may have different degrees of affinity to the different types of resources and task-to-resource mapping becomes more complex issue.

2.2.1 Resource Constraint

A resource can be denoted as a structure that can be used by a task process to continue its execution. Typically, a resource could be a set of variables, a memory range, a program, or registers or a data structure. A dedicated resource for particular task is said to be private resource, whereas a resource that can be accessed by multiple tasks is known as shared resource.

A task which requires an exclusive resource but the resource is not immediately available, then the task is said to be blocked task. All the blocked tasks are stored in a queue, When a *running* task requires an exclusive resource, it goes into the *waiting* state. When the required resource becomes free, a free signal is received by the *waiting* tasks. Then a task leaves the waiting state and moves to the ready state and the scheduling algorithm allocates the highest-priority task to the processor. The transition diagram for the above scenario is depicted above in Fig. 2.2.

2.2.2 Metrics for Scheduling Evaluation

Different schedule evaluation criteria have evolved based on what the scheduler intends to optimize and each such objective requires a different solution technique. A few important evaluation metrics are given below:

- Deadline assurance: For hard real-time tasks, the deadline should be satisfied by completing the task before deadline.
- Late and early completion of tasks: The time difference between the actual and stipulated completion-time of the task. If a task completes after its deadline, the evaluation criterion is called tardiness. Similarly, the evaluation criterion is called earliness, if the task finishes before deadline.

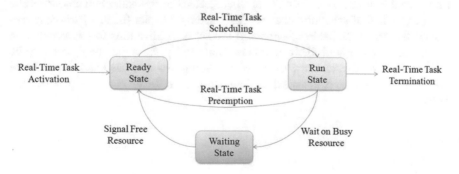

Fig. 2.2 Resource constraint

- Fairness of each task: This criteria measures the rate of execution of a task relative to other tasks in the system. Two tasks possessing equal priority values are said to be scheduled with equal fairness only if they are executed at the same rate.
- Scheduling overhead: Scheduling overhead is incurred by the scheduler to select the next task to be allocated to processor. The analysis of this overhead is necessary if the scheduling requires to be done dynamically at run time.
- Context-Switch Overheads: Context-Switches are caused due to task preemptions. The increased use of multiprocessor and multi core processors in today's computing systems have made the number of context-switches within the length of the schedule an important schedule evaluation criteria. It has to be noted that for FPGAs these context-switches overhead could be of the order of few hundred milliseconds and thus, need to be handled judiciously.
- Other Criteria: Energy Consumption, Fault Tolerance, etc.: With the advent of different application types and newer types of architectural platforms, newer schedule evaluation metrics have emerged. For example, in power constrained embedded systems, schedules that minimize total energy consumption using mechanisms like Dynamic Voltage Scaling (DVS) and Dynamic Power Management (DPM) are very important. Again, hard real-time systems like automotive and avionics systems often demand schedules that remain valid even in the presence of system faults.

2.3   Real-Time Scheduling

Scheduling appears in any domain where there is a need to allocate limited available resources in order to serve a certain number of tasks. It is then necessary to coordinate the use of such resources so that the tasks may run to completion as efficiently as possible. This efficiency means optimizing one or many criteria. Such criteria could be to minimize the schedule length (makespan), maximize resources utilization, minimize the number of tasks that must be rejected due to insufficient resources etc.

The problem of scheduling can be described by a triplet $\{\alpha, \beta, \gamma\}$ where α represents the set of available resources, β the set of applications to be executed on the resources along with their time constraints and γ the objective function to be optimized. The generic multi-resource scheduling problem has been shown to be NP complete and hence, many scheduling heuristics of lesser complexity have been proposed.

2.3.1 Offline Versus Online Scheduling

The real-time tasks must ensure its deadline. In order to achieve this, two different scheduling approaches are available i.e. offline and online scheduling.

In offline scheduling, the scheduler has the prior information related to the task i.e. arrival times, execution times, precedence constraints, etc. With these information, schedule is generated at design time and at runtime, execution begins as per the generated schedule. Offline scheduling is also called as static scheduling [CMM03]. Offline scheduling is beneficial to find the optimal allocation of tasks.

On the other hand, in case of online scheduling, the task's information is not known in advance. Online scheduling algorithms make their scheduling decisions at runtime based on information about the arrived tasks. Online scheduling is flexible than offline scheduling as it can be employed where arrival pattern of tasks dynamically changes. However, they may incur significant overheads because of runtime processing. Usually, online scheduling algorithms try to produce a better solution, but cannot guarantee the optimal solution. Online scheduling is also referred as dynamic scheduling.

Hard real-time tasks demand highly predictable behavior. Before the task execution, the admission controller should assure that all tasks will be scheduled within the deadline. In order to derive a feasible schedule, the system considers worst-case execution values. In static real-time systems, the fixed task set is known design time. Thus it is feasible to generate the schedule in offline and store the entire schedule in a table. At runtime, a dispatcher will assume the responsibility to select the task from the table and will put it in the running state. The main advantage of such offline algorithm is that it does not impose high runtime overheads. On the other hand, for the dynamic real-time systems, tasks may arrive at runtime; hence the online scheduling algorithm must guarantee that all the accepted tasks will be feasibly scheduled. Figure 2.3 shows the typical method adopted in dynamic real-time system to guarantee the feasible schedule of dynamically arrived tasks.

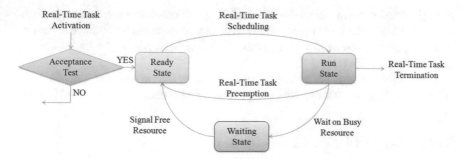

Fig. 2.3 Guarantee mechanism for dynamic real-time system

2.3.2 Real-Time Scheduling for Uniprocessor Systems

Scheduling theory has been intensively studied over the years and uniprocessor scheduling takes the lion's share in the rich literature review. Many optimal uniprocessor scheduling algorithms have been proposed along with their schedulability analysis. Here below are some scheduling algorithms.

2.3.2.1 Rate Monotonic (RM)

The RM algorithm [LL73] is a preemptive *static priority* scheduling scheme for periodic and independent tasks systems. In this scheme, tasks are assigned an integer priority value that remains fixed. The tasks with higher priority always preempts the task with lower priority. In RM, shorter the period of a task, higher becomes its priority. Liu and Layland (1973) [LL73] have proven RM to be an optimal static priority scheduler for preemptive task systems.

An important shortcoming of the RM algorithm (as shown by Liu and Layland) is that even on uniprocessor systems no more than 69% of the processor may be utilized to ensure scheduling feasibility of a set of tasks under rate-monotonic priority assignment, in the worst case.

2.3.2.2 Deadline Monotonic (DM)

DM [Alt95] is a *static priority* scheduling algorithm that gives the highest priority to the task with the least relative deadline d_i. DM could be used with periodic, aperiodic and sporadic tasks systems.

2.3.2.3 Earliest Deadline First (EDF)

EDF [Liu00a] is a *dynamic priority* scheduling scheme where the task with closest deadline, obtains the higher priority. Priorities are reassessed and updated at runtime if necessary (e.g. on each task arrival). EDF has been proven to be optimal for preemptive periodic tasks.

Later, EDF has also been shown to be optimal in the case of non-periodic tasks. EDF scheduling outperforms RM and produces less preemption compared to RM.

2.3.2.4 Least Laxity First (LLF)

Another *dynamic priority* scheduling is the Least Laxity First (LLF) [Liu00a] scheduler. In LLF, at each scheduling point, a laxity value (L) is computed for each task and the task having the smallest laxity is assigned the highest priority. Here, the laxity value (L_i) for a task T_i is defined as: $L_i = d_i - (t - re_i)$, where d_i denotes the deadline of T_i, t the current time and re_i the remaining execution requirement of T_i.

However, in LLF algorithm, many tasks are likely to have the same laxity (which means same priority), leading to a situation where many preemptions are performed in a short time interval, which is not desirable.

2.3.3 Processor Utilization

Given a set of n real-time tasks, the "*processor utilization*" can be defined as the fraction of CPU time spent in executing a task. Let us assume, e_i is the worst-case execution requirement of a task and p_i is the period/ deadline. Now, $\frac{e_i}{p_i}$ denotes the fraction of CPU time spent in executing T_i, the "*processor utilization (PU)*" for n tasks can be defined as

$$PU = \sum_{i=1}^{n} \frac{e_i}{p_i}$$

The PU provides a measure of the load on the processor due to the real-time task set. From the above equation, it can be noted that the PU can be improved by increasing e_i or by decreasing p_i. However, there exists a maximum value of PU which decides whether a given task set is schedulable or not, as discussed later in this chapter.

2.3.4 Real-Time Scheduling for Multiprocessor Systems

An increasing number of real-time systems require more than one processors to achieve their performance goals. The problem of scheduling tasks on multiple processors cannot be seen as a simple extension of the uniprocessor scheduling due to additional constraints such as: task migration overheads, inter-task communication overheads etc. Traditionally, there are two classes of scheduling algorithms for multiprocessor platforms: partitioned scheduling and global scheduling.

2.3.4.1 Global Versus Partitioning Scheduling

In global scheduling, all ready tasks are stored in a single priority queue among which scheduler selects the highest priority task at each invocation irrespective of which processor is being scheduled. In a purely partitioned approach on the other hand, the set of tasks is partitioned into as many disjoint subsets as there are processors available, and each such subset is assigned to a distinct processor [CFH+04]. After this mapping is obtained, all instances/jobs of a given task are executed only on the processor to which it is associated.

Due to the allowance of task migrations, global scheduling methodologies are typically able to achieve higher schedulability compared to partitioned scheduling. However, task migrations and preemptions come at the cost of increased runtime overheads. Therefore attempts have been made to devise schemes which are able to restrict migrations while satisfying the performance goals.

Pros and Cons

The main advantage of partitioning is that it allows the multiprocessor scheduling problem to be reduced to a set of uniprocessor ones. Within each processor, a separate well known uniprocessor scheduler like Rate Monotonic (RM), Earliest Deadline First (EDF), etc. may be easily applied. In addition, the overhead of inter-processor task migrations is smaller than global scheduling. Finally, because task-to-processor mapping (which task to schedule on which processor) need not be decided globally at each scheduling event, the scheduling overhead associated with a partitioned strategy is lower than that associated with a global strategy [AT06, CFH+04]. However, partition based scheduling approaches may often be plagued by low resource utilization. Oh et al. in [OB98] showed that on homogeneous multiprocessor systems where no task migration between processors is allowed and each processor schedules tasks preemptively employing the Rate Monotonic policy, the maximum utilization that may be achieved is just 41%. When EDF, a well known optimal scheduler for uniprocessor systems, is applied to multiprocessors using a fully partitioned approach disallowing task migrations, the worst case utilization bound reduces to 50% [LGDG00].

2.3.4.2 Recent Trends in Real-Time Multiprocessor Scheduling

Recent multiprocessor scheduling techniques like ERfair [AS00] and DP-Fair [LFS+] reveal the fact that optimal resource utilization can be achieved irrespective of the skewness in task weights/periods. These "fair" scheduling ensures within a time span all tasks will be executed by same rate which is termed as "proportional fair". However, such proportional fairness is achieved at the cost of higher preemptions/migrations (moving tasks from one processor to another).

ERfair Scheduler: ERfair schedulers mandate execution of each task T_i to proceed proportionally at a rate lower bounded by a parameter called its *weight* (wt_i) which is defined as the ratio of its execution requirement (e_i) and period/deadline (p_i). To maintain ERfairness at the end of any given scheduling time slot t, $s_i \leq t \leq s_i + p_i$, at least $\frac{e_i}{p_i} \times (t - s_i)$ of total execution requirement of e_i must be completed for each task T_i, where s_i is the start time of task T_i. ERfair ensures schedulability if the summation of weights of the n tasks is atmost the number of processors m ($\sum_{i=1}^{n} e_i/p_i \leq m$).

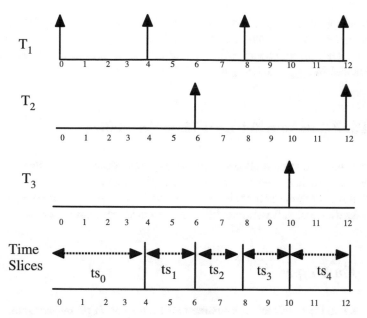

Fig. 2.4 Time slices in DP-Fair

DP-Fair Scheduler: Although ERfair is an optimal scheduler allowing full resource utilization, it suffers from high scheduling overheads as well as inter-processor task migration overheads. Recently Deadline Partitioning-Fair (DP-Fair), a lower overhead optimal scheduling strategy with a more relaxed proportional fairness constraint has been proposed. DP-Fair sub-divides time into slices/chunks demarcated by the deadlines of all tasks. Time slice ts_i denotes the interval between the $(i-1)$th and ith task deadlines. For example, as shown in Fig. 2.4, let us assume three ready periodic tasks T_1, T_2, T_3 at time 0 with periods/deadlines 4, 6, 10, respectively.; therefore, there are distinct deadlines at 4, 6, 8, 10 and 12. Hence, we have five distinct time slices ts_1, ts_2, ts_3 ts_4 and ts_5.

Within each time slice, each task is assigned a "*per-slice workload*" which is equal to its weight times the length of the time slice. If tsl_i be the length of the ith time slice, then task T_i should complete its allocated workload ($\frac{e_i}{p_i} \times tsl_i$) within the time slice. Although the exact intra-time slice task scheduling policy may vary, any adopted strategy must obey the following three rules:

1. Laxity of the executing tasks should be zero.[1]
2. Never run a task with no workload remaining in the slice
3. The idle processor time can be maximum of

$$(m - \sum \frac{e_i}{p_i}) \times (length\ of\ time\ slice)$$

---

[1] Laxity is the maximum amount of time a task can wait and still meets it deadline.

DP-Fair is optimal multiprocessor algorithm. The crux idea behind DP-Fair is that DP-Fair maintains ERfairness constraint only at task period/deadline boundaries. Obviously, DP-Fair will be an optimal algorithm for m processors, if the following condition satisfies $\sum_{i=1}^{n} e_i/p_i \leq m$ i.e. $PU \leq m$.

2.4   Fault Tolerance for Real-Time Scheduling

Fault-tolerance techniques allows systems to operate despite of hardware and software faults. Specifically, for real-time systems, the system designer must ensure that the probability of system failure is small. In a real-time system, there is a probability that the hardware or software may fail which in turn jeopardize the deadlines.

2.4.1   Fault Types

There are two types of faults: Permanent and transient. A permanent fault remains in the system until it is repaired or replaced. Transient faults appears in the system at a certain instant and disturbs the system however, it does not sustain in the system for long time,

2.4.2   Fault Detection

In order to employ fault tolerance technique, at first, transient faults have to be detected. A few popular detection techniques are:

Watchdogs. Watchdog programs periodically check transmitted data for the presence of errors. Similarly, watchdog timer, monitors whether the execution of a task finishes within the deadline.

Sanity check In this method, a desired output is compared with the current output. Any mismatches indicates the existence of a fault.

2.4.3   Fault Tolerance Techniques

- TMR (Triple Modular Redundancy) In this technique, multiple copies of a program are executed on different processors. The correct output is generated by conducting a majority voting. If any one of the three processors fails then the other two can mask and mitigate the results.

- PB (Primary/Backup) in this technique, each task is equipped with two versions, i.e. a primary version and backup version. In the processor, initially the primary version is executed, if any error is found at the end of its execution then the backup version is re executed to mitigate the fault.

2.4.4 Fault Tolerance Scheduling

In fault-tolerance scheduling fault is typically handled by re-executing tasks on a backup processor to deliver the correct result [GZA+17]. However, such re-execution of tasks is energy consuming and it has to be noted that energy is an important factors for real-time systems , as these devices performs under limited power source such as batteries [GZA+17]. Now, energy-aware execution of tasks can be tackled using two main techniques i.e. (i) *Dynamic Voltage Scaling (DVFS)* technique which trade offs between processor speed and power dissipation [RAZ19] and (ii) *Dynamic Power Management (DPM)*, which keeps idle processor and components in low-power sleep states and hence, preserves power [NDS18]. Recently, heterogeneous multicore systems are becoming widely popular for real-time systems. In such systems, processing elements have different power/performance and judicious use of such processors, improves energy efficiency of the system. ARM's big little systems, Xilinx ZYNQ platform represents such heterogeneous systems [MSC17, ZCC+17].

Recent real-time scheduling strategies [HAZ16, FHY17, ZAZ12], have combined the problem of energy minimisation along with fault tolerance techniques. However, these techniques are mainly devised either for uniprocessor systems or for homogeneous multiprocessors. In case of heterogeneous systems, standby sparing and primary/backup techniques [GZA+17, RAZ17, NDS18], are the contemporary solutions. All of these techniques employed standard scheduling scheme like Earliest-Deadline- First (EDF) and Earliest-Deadline-Late (EDL) scheduling policies. Moreover, these works assumes that tasks have a fixed and common deadline. However, In modern real-time systems, such assumption is bit strait and generally no longer valid.

2.5 Imprecise Computation Based Real-Time Task

Another recent trends in real-time computing is to employ imprecise computation (IC) task. An IC task, T_i is logically decomposed into a mandatory part with execution requirement of M_i and an optional part with execution requirement of O_i. Each IC task has to execute M_i units in order to generate an acceptable result. Optional part O_i can only be started after mandatory part M_i completes. Hence, the entire portion of the mandatory part has to be executed for generating initial acceptable result, but the optional part will attempt to improve the the accuracy of the initial result. O_i can be executed partially/completely based upon the available resources. The execution length l_i of an IC task T_i can be defined as :

$$l_i = M_i + O_i \times \mu \tag{2.1}$$

where, μ denotes the fraction of the optional part, which is being executed and varies between 0 and 1. Thus, $\mu = 1$ denotes complete execution of O_i units and maximum possible accurate result.

As mentioned earlier, in real-time embedded systems, energy minimization is a key requirement. A significant research efforts can be observed towards minimizing the energy for tasks for multi-processor systems [NHG+16]. Recently in [WZC+17], the authors introduce the concept of IC task to handle the limitations of energy of a real-time system. This strategy accepts approximate results, if the energy budget is limited and carries out further computations to gain accurate results, if available energy is sufficient. This work is an attempt to introduce the concept of "imprecise computation" to handle energy constraint. However this technique motivates many researchers to tackle the problem of energy efficient and fault tolerant real-time scheduling by employing imprecise computation.

A few researchers have explored IC tasks scheduling for embedded real-time systems with energy savings, e.g., [WZC+17, MDOSZ17, ZYW+17]. In [YVH08], Yu et al. introduced "IC", where tasks have mandatory portion and an optional portion. The authors then proposed a "dynamic-slack-reclamation" technique to improve the overall system accuracy and make the system energy efficient. In [SK10], the authors compared the performance of exsiting real-time scheduling schemes like Highest Level First (HLF) and Least Space Time First (LSTF) between set of tasks, where a fixed set of tasks employed imprecise computation and the other did not. However, this work did also not consider energy limitations.

Energy aware scheduling of IC tasks are also considered in [MKS18]. Here, the authors employed DVFS based technique to make the scheduling strategy energy efficient. However, as DVFS decreases the supply voltage and frequency to save power, the fault rate of the system can significantly increase [HAZ16], and thus reduce the system's reliability. In order to circumvent the problem of DVFS technique, in [XP14] Xiang et al. propose the idea of finding "energy efficient frequency" in a multi-core platform.

2.6   Real-Time Scheduling on FPGA

Let us consider an use-case where executing real-time task sets on FPGAs may be very useful. Assume that FPGAs are employed as back up platform for real-time tasks in a complex safety-critical system. These FPGAs will be activated as backups whenever the primary processors would fail. In such scenario, FPGAs will be responsible for executing the real-time tasks that were previously running on the failed processors. Therefore, a well defined scheduling methodologies are essential.

To ensure feasible schedule on backup FPGAs, feasibility criteria and admission control strategies for real-time tasks shold be checked. Scheduling of real-time tasks on FPGAs with efficient resource utilization is a challenging problem. The complexity of real-time scheduling on reconfigurable platforms originates from the inherent architectural constraints in FPGAs which impose a few unique hurdles that any resource allocation technique must overcome.

2.6.1 Challenges for FPGA-Based Scheduling

- The first challenge is to employ a efficient strategy which consumes less overhead but achieves high spatial parallelism. A task in FPGA typically consumes a rectangular subregion and may be placed anywhere of the FPGA floor. Parallel execution of a given set of tasks (regions in FPGAs) is achieved by placing the tasks on the FPGA such that no task region overlaps with the FPGA boundaries or with other placed tasks. This is unlike general purpose multiprocessors where parallelism is enforced through simultaneous execution of the tasks on physically distinct processing elements. This makes the problem of maximizing spatial parallelism through effective placement NP-complete in the strong sense [CCP06].
- The problem of effective placement has been further complicated by the advent of heterogeneous reconfigurable platforms in the recent past. In these platforms, the FPGA may contain hard embedded blocks along with CLBs. These heterogeneous hard components may include of BlockRAMs (BRAMs), Multipliers (MULs), Digital signal processors (DSPs). These unique hard blocks helps to speed-up task's execution. However, these HBs impose additional placement constraint for a task which requires these special elements.
- As we discussed before, in order to achieve high resource utilisation a preemptive scheduling strategy is beneficial. The third hurdle is to design effective preemptive scheduling mechanisms in the face of significant context switching overheads. Context switches in FPGAs are actually effected through hardware reconfigurations. Each context switch in FPGA will consume in the range of a few or even higher than ten milliseconds. Hence, context switches must be judiciously used and otherwise, it will impose additional overhead which will lea to deadline misses. On the other hand, for CPUS, context-switches come with low overheads (in the order of hundreds of nano seconds). Thus, the scheduling algorithms for these systems do not consider overheads related to preemptions and migrations.

2.6.2 Preemption of Hardware Tasks

A non-preemptive task only releases the CPU voluntarily. On the other hand, preemptive tasks can be interrupted (to let another task execute) and resumed later. One advantage of preemptive scheduling is its capability to schedule tasks with widely different periods.

For example, let us consider a typical periodic task T_1, from the domain of automotive control systems, having an execution time of 10 ms and period of 100 ms. Even if this task is always assigned the highest priority in a non-preemptive system, execution time of all other tasks must not be greater than 90 ms to ensure that no instance of T_1 ever misses its deadline in the worst case. This enforces manual partitioning of tasks having large execution times into small subtasks, making the design and implementation of the system very complex. Another advantage of preemptive scheduling is the possibility of admission control using utilization bounds and achieving an efficient resource utilization, up to 100% for a preemptive scheduler like Earliest Deadline First (EDF) [Liu00a].

Though preemptive scheduling provides flexibility and higher resource utilization, yet there has been very little work in this regard for re-configurable platforms. This is due to the challenges involved in saving the state of a partially completed hardware task and restoring saved states to re-initiate execution. The "contexts" of a hardware task are usually stored in elements like Flip-flops,LUT-RAMs of CLBs, BRAMs and MULs etc. Context switching in FPGA involves following activities:

1. Capture the contexts of tasks that were executing prior to a switch.
2. Update he contexts and save them in an external memory.
3. Form a new bitstream comprising of tasks that will execute after switching.
4. Download the new bitstream so that tasks can start their execution after preemption.

Authors in [JTW07] have employed a mechanism called *Scan Path* generation to allow task context extraction/insertion on re-configurable platforms. However, to enable hardware preemption, the methodology necessitates addition of task specific components known as scan-chains to hardware tasks, thus incurring significant spatial overheads. An improvement over this strategy is the *bitstream readback* methodology [KP05]. In [KP05], the authors proposed hardware context switching on Virtex-4 FPGAs using bitstream read-back through ICAP along with some additional combinational logic inside the re-configurable region.

Now, traditional technologies for context switch, which required to save entire task bitstreams, incurred about ≈ 7 ms to extract the context of a task having size ≈ 500 KB [KP05]. Needless to say, that such extraction time is unaffordably high. However, recent literature [JTHT, JTE+12] has discussed that the instantaneous state of a task is contained within at most $\approx 8\%$ of its entire bitstream and the extraction of this part of the bitstream during context switch will be sufficient. Through such *selective bitstream read-back* scheme, context switch overheads between tasks of typical sizes (≈ 500 KB) may be drastically reduced to only $\approx 700\ \mu$ s [JTE+12] using this intelligent selective extraction mechanism. Thus, the actual extraction overhead may be reduced to atmost 1/10th of the full bitstream read-back time. Hence, the overall context switch overhead will also be reduced. A detailed discussion on the quantification of hardware context switching overheads for a specific FPGA family can be found in next section.

2.6.3 Existing FPGA-based Real-Time Scheduling Techniques

Diessel and Elgindy [DE01], combined placement along with a temporal scheduling mechanism for non real-time preemptible task sets. Tasks are allocated from a ready queue starting from the bottom-left position using a first-fit strategy. When the allocator fails to accommodate the next pending task, it attempts to create additional space by preempting and locally repacking the currently running tasks utilizing spare logic resources created due to the partial completion of execution of these tasks.

Spatio-temporal scheduling of tasks with pre-assigned priorities have also been considered for reconfigurable systems, where priority could be based on the temporal (deadline, laxity) or geometrical (size, aspect-ratio) properties of the tasks (or both) [WL12]. Walder and Platzner [WP03] used non-preemptive online scheduling schemes like *First Come First Serve (FCFS) and Shortest Job first (SJF)* for block-partitioned 1D reconfigurable devices. In [CH05], authors present a scheduling mechanism called *Classified Stuffing (CS)*, where both geometrical and temporal parameters have been used to obtain priorities. The tasks are first inserted into a ready queue based on their temporal priorities (For non real-time tasks, these priorities are obtained using the *Shortest Remaining Processing Time (SRPT)* strategy, while for real-time tasks, the *Least Laxity First (LLF)* strategy is followed). The tasks in the ordered ready queue are then placed on the basis of a spatio-temporal parameter called *Space Utilization Rate (SUR)*, which is defined as the ratio between the area requirement and the execution requirement of a task.

Due to the challenges involved in the hardware tasks preemption on FPGAs, non-preemptive scheduling strategies was an alternate. In [GKK15], authors presented, Clairvoyant EDF [Eke06], a non-preemptive real-time task scheduling and placement using. However, these real-time non-preemptive scheduling approaches restrict resource utilization [Lau00, DB11]. Although there exist works which have employed preemptive scheduling techniques for FPGAs. Danne and Platzner [DP05] consider the problem of scheduling preemptive periodic real-time tasks on FPGAs by using a flexible 2D area model. They proposed the EDF-Next Fit (EDF-NF) algorithm, a variant of EDF [Liu00a]. However, being based on EDF, this algorithm only achieves a resource utilisation Of 50%.

To achieve the higher resource utilization and minimize task rejections, a different dynamic scheduling and placement approach has been discussed in [Mar14]. In this article, the author assumes that a hardware task could have multiple hardware variants (varying size) such that, larger the size of a task, faster is its performance. Selection of appropriate hardware task variants is a trade-off between maximal utilization of reconfigurable resources versus the timing requirements of the tasks. A similar algorithm which considers multiple hardware task shapes has been proposed in [WBL+14]. The authors showed that resource utilizations using conventional scheduling algorithms like EDF and LLF may be significantly improved by using the flexibility of multiple shapes, against a rigid task scenario. This work neglected reconfiguration overheads and assumed tasks to be soft real-time in nature. However, reconfiguration overhead is a major constraint which may adversely affect the temporal performance of tasks if not handled appropriately. In [LMBG09], authors described the *reuse and partial reuse* approach which allows a single configured task on the reconfigurable floor to be shared and reused among multiple applications, thus reducing the number of reconfigurations required.

2.6.4 Real-Time Preemptive Scheduling: Uniprocessors Versus Multiprocessors Versus FPGAs

Preemption is an important technique that allows real-time systems to achieve high resource utilization by lending it the flexibility to co-schedule tasks having different behaviors (in terms of execution times, periodicity, task recurrence etc.) in varying execution environments (ranging from uniprocessors, homogeneous/heterogeneous multiprocessors and even reconfigurable processing cores). Non-preemptive tasks are characterized by the fact that once allocated a processor at the beginning of execution, the processor cannot be relinquished from the task until its completion. Preemptive tasks on the other hand, can be interrupted (to possibly allow other tasks to execute) before completion and resumed later. An important advantage of preemptive scheduling is its capability to schedule tasks with widely varying periods. For example, let us consider a typical periodic task T_1, from the domain of automotive control systems, having an execution time of 10 ms and period of 100 ms. Even if this task is always assigned the highest priority in a non-preemptive system, execution time of all other tasks must not be greater than 90 ms to ensure that no instance of T_1 ever misses its deadline in the worst case. To ensure this, tasks having large computation times must be manually partitioned into small sub-tasks, and this makes the scheduler design very tedious especially in large and complex systems. In comparison, by allowing preemption, dynamic unicore scheduling strategies like Earliest Deadline First (EDF) [Liu00a] achieves full resource utilization without imposing any restriction on task execution times and/or deadlines/periods. These preemptive systems are also practically realizable in most cases because context switching overheads involved in each preemption is typically very low and may usually be neglected.

So in a nutshell, non-preemptive scheduling approaches restrict resource utilization especially in scenarios where the individual task periods are skewed [Lau00]. The situation becomes worse as we shift from unicore to multicore systems with numerous processing elements [DB11]. Traditionally, there have been two principal approaches towards real-time scheduling of tasks on multicore systems: global scheduling and partitioned scheduling [AJ00]. In fully partitioned scheduling, all tasks are first assigned to dedicated cores. Tasks allocated to a given core are executed on that core until completion. The main advantage of partitioned scheduling is that the multicore scheduling problem can be amicably reduced to a set of unicore problems and scheduled using well-known approaches like EDF. However, a critical drawback of this approach is that even by employing preemptive scheduling approaches like EDF (which offer 100% resource utilization on unicore systems), not more than 50% of the system capacity may be utilized in the worst-case [AJ03].

In global scheduling on the other hand, all tasks are maintained in a single ready queue. At each scheduling event, the m (denotes the number of cores) highest priority tasks are selected from this queue. By allowing the flexibility of inter-core task migrations during execution, global schedulers can typically achieve significantly superior resource utilization compared to partitioned schemes. Pfair and its work conserving variant ERfair [AS00], both fully global schemes, were the first multicore scheduling strategies that allowed optimal resource utilization irrespective of the skewness in task weights/periods. This optimality was achieved by maintaining proportional fair execution progress for all tasks at each time instant throughout the length of the schedule. However, in order to provide such accurate proportional fairness, these schemes incur unrestricted migrations and preemptions which leads to high context switch related overheads even in closely-coupled multicores with shared caches.

The need to control context switch overheads led to the design of more recent semi-partitioned approaches like DP-Fair and Bfair. These approaches partition time into slices/chunks, demarcated by the event of task arrivals and departures. In eachh time slice, each task is allocated a work load equal to its proportional fair share. The task shares within each core are usually scheduled using EDF-like strategies and completed by the end of the time slice. By deviating from the need to maintain strict proportional fairness at all times, these strategies are able to guarantee resource utilization optimality while incurring atmost $m - 1$ migrations within time slices. Thus, context switch overheads in these semi-partitioned schemes are significantly reduced in comparison to global approaches like ERfair. However, although the semi-partitioned DP-Fair scheme may be considered a prominent state-of-the-art scheduler for dynamically arriving task sets in multi-core systems, it cannot be directly employed in platforms such as FPGAs. This is due to the inherent architectural constraints in FPGAs which lead to non-negligible reconfiguration overheads in the order of a few to tens of milliseconds [HSH09]. Thus, in case of FPGAs, context switches come at a premium and they must be judiciously handled as well as have to be correctly accounted for within any given interval of time. Otherwise, correct estimation of available system capacity will not be possible and this may lead to huge task deadline misses. Due to this fact, algorithms for general purpose multicore systems are bound to fare poorly on FPGAs.

2.7 Conclusions

Real-time systems have a dual notion of correctness - logical and temporal. Scheduling problem deals with the temporal correctness. The chapter discusses real-time scheduling algorithms for uniprocessor and multiprocessor systems. The limitations of existing scheduling methodologies on multiprocessor systems have been discussed. This chapter also introduces fair scheduling approaches like ER-fair, DP-Fair etc., where the scheduling scheme ensures ta hat each task must make progress according to a predetermined rate. Fault tolerance scheduling along with the concept of contemporary "imprecise computation" technique has been discussed. It is observed that with the increasing complexity of real-time systems, co-scheduling of periodic and aperiodic real-time tasks is a challenging topic.

FPGAs are becoming a popular choice for many of todays real-time systems. However, devising periodic real-time scheduling strategies for FPGAs by directly employing existing techniques (which are targeted towards general purpose multi-processors, like DP-Fair) is a challenging problem. The first challenge is to devise a low overhead but efficient placement strategy so that the spatial parallelism could be maximised. Secondly, preemption or context switching can not be easily employed in the FPGAs. In the next chapter, we discuss two novel online real-time scheduling strategies for fully and partially reconfigurable systems, respectively.

References

[AJ00]   B. Andersson, J. Jonsson, Some insights on fixed-priority preemptive non-partitioned multiprocessor scheduling, in *Proceedings of the IEEE Real-Time Systems Symposium, Work-in-Progress Session* (2000)

[AJ03]   B. Andersson, J. Jonsson, The utilization bounds of partitioned and pfair static-priority scheduling on multiprocessors are 50%, in *Real-Time Systems, 2003. Proceedings. 15th Euromicro Conference on*, (IEEE, 2003), pp. 33–40

[Alt95]   P. Altenbernd, Deadline-monotonic software scheduling for the co-synthesis of parallel hard real-time systems, in *Proceedings of the 1995 European conference on Design and Test* (IEEE Computer Society, 1995), p. 190

[AS00]   J.H. Anderson, A. Srinivasan, Early-release fair scheduling, in *Real-Time Systems, 2000. Euromicro RTS 2000. 12th Euromicro Conference on* (2000), pp. 35–43

[AT06]   B. Andersson, E. Tovar, Multiprocessor scheduling with few preemptions, in *Embedded and Real-Time Computing Systems and Applications, 2006. Proceedings. 12th IEEE International Conference on* (IEEE, 2006), pp. 322–334

[CCP06]   D. Chen, J. Cong, P. Pan, Fpga design automation: A survey. Found. Trends Electron. Des. Autom. **1**(3), 139–169 (2006)

[CFH+04]   J. Carpenter, S. Funk, P. Holman, A. Srinivasan, J.H. Anderson, S.K. Baruah, A categorization of real-time multiprocessor scheduling problems and algorithms (2004)

[CH05]   Y.-H. Chen, P.-A. Hsiung, Hardware task scheduling and placement in operating systems for dynamically reconfigurable soc, in *Embedded and Ubiquitous Computing–EUC 2005* (Springer, 2005), pp. 489–498

[CMM03]   R.W. Conway, W.L. Maxwell, L.W. Miller, *Theory of scheduling* (Courier Corporation, 2003)

[DB11]   R.I. Davis, A. Burns, A survey of hard real-time scheduling for multiprocessor systems. ACM Comput. Surv. (CSUR) **43**(4), 35 (2011)

[DE01]   O. Diessel, H. Elgindy, On dynamic task scheduling for fpga-based systems. Int. J. Found. Comput. Sci. **12**(05), 645–669 (2001)

[DP05]   K. Danne, M. Platzner, Periodic real-time scheduling for fpga computers, in *Intelligent Solutions in Embedded Systems, 2005. Third International Workshop on* (2005), pp. 117–127

[Eke06]   C. Ekelin, Clairvoyant non-preemptive edf scheduling, in *Real-Time Systems, 2006. 18th Euromicro Conference on* (IEEE, 2006), p. 7

[FHY17]   M. Fan, Q. Han, X. Yang, Energy minimization for on-line real-time scheduling with reliability awareness. J. Syst. Softw. **127**, 168–176 (2017)

[GKK15]   Z. Guettatfi, O. Kermia, A. Khouas, Over effective hard real-time hardware tasks scheduling and allocation, in *Field Programmable Logic and Applications (FPL), 2015 25th International Conference on* (IEEE, 2015), pp. 1–2

[GZA+17]   Y. Guo, D. Zhu, H. Aydin, J.-J. Han, L.T. Yang, Exploiting primary/backup mechanism for energy efficiency in dependable real-time systems. J. Syst. Arch. **78**, 68–80 (2017)

[HAZ16]   M.A. Haque, H. Aydin, D. Zhu, On reliability management of energy-aware real-time systems through task replication. IEEE Trans. Parallel Distrib. Syst. **28**(3), 813–825 (2016)

[HSH09]   P.-A. Hsiung, M.D. Santambrogio, C.-H. Huang, *Reconfigurable System Design and Verification* (CRC Press, 2009)

[JTE+12]   K. Jozwik, H. Tomiyama, M. Edahiro, S. Honda, H. Takada, Comparison of preemption schemes for partially reconfigurable fpgas. IEEE Embedded Syst. Lett. **4**(2), 45–48 (2012)

[JTHT]   K. Jozwik, H. Tomiyama, S. Honda, H. Takada, A novel mechanism for effective hardware task preemption in dynamically reconfigurable systems. FPL 352–355 (2010)

[JTW07]   S. Jovanovic, C. Tanougast, S. Weber, A hardware preemptive multitasking mechanism based on scan-path register structure for fpga-based reconfigurable systems, in *Adaptive Hardware and Systems, 2007. AHS 2007. Second NASA/ESA Conference on* (IEEE, 2007), pp. 358–364

[KP05]   H. Kalte, M. Porrmann, Context saving and restoring for multitasking in reconfigurable systems, in *Field Programmable Logic and Applications, 2005. International Conference on* (IEEE, 2005), pp. 223–228

[Lau00]   S.M. Lauzac, *On multiprocessor scheduling of preemptive periodic real-time tasks with error recovery*. PhD thesis, University of Pittsburgh (2000)

[LFS+]   G. Levin, S. Funk, C. Sadowski, I Pye, S. Brandt, *Dp-fair: A Simple Model for Understanding Optimal Multiprocessor Scheduling* (ECRTS 2010)

[LGDG00]   J. Maria López, M. García, J. Luis Diaz, D.F. Garcia, Worst-case utilization bound for edf scheduling on real-time multiprocessor systems, in *Real-Time Systems, 2000. Euromicro RTS 2000. 12th Euromicro Conference on* (IEEE, 2000), pp. 25–33

[Liu00a]   J.W.S. Liu, *Real-Time Systems* (Prentice Hall, 1st edition, 2000)

[LL73]   C.L. Liu, J.W. Layland, Scheduling algorithms for multiprogramming in a hard-real-time environment. J. ACM (JACM) **20**(1), 46–61 (1973)

[LMBG09]   Y. Lu, T. Marconi, K. Bertels, G. Gaydadjiev, Online task scheduling for the fpga-based partially reconfigurable systems, in *Reconfigurable Computing: Architectures, Tools and Applications* (Springer, 2009), pp. 216–230

[Mar14]   T. Marconi, Online scheduling and placement of hardware tasks with multiple variants on dynamically reconfigurable field-programmable gate arrays. Comput. Electr. Eng. **40**(4), 1215–1237 (2014)

[MDOSZ17] I. Méndez-Díaz, J. Orozco, R. Santos, P. Zabala, Energy-aware scheduling mandatory/optional tasks in multicore real-time systems. Int. Trans. Oper. Res. **24**(1–2), 173–198 (2017)

[MKS18]   L. Mo, A. Kritikakou, O. Sentieys, Energy-quality-time optimized task mapping on dvfs-enabled multicores. IEEE Trans. Comput.-Aided Des. Integr. Circ. Syst. **37**(11), 2428–2439 (2018)

[MSC17]   A. Majumder, S. Saha, A. Chakrabarti, Task allocation strategies for fpga based heterogeneous system on chip, in *IFIP International Conference on Computer Information Systems and Industrial Management* (Springer, 2017), pp. 341–353

[NDS18]   P.P. Nair, R. Devaraj, A. Sarkar, Fest: Fault-tolerant energy-aware scheduling on two-core heterogeneous platform, in *2018 8th International Symposium on Embedded Computing and System Design (ISED)* (IEEE, 2018), pp. 63–68

[NHG+16]   S. Narayana, P. Huang, G. Giannopoulou, Lothar Thiele, and R Venkatesha Prasad. Exploring energy saving for mixed-criticality systems on multi-cores. In *2016 IEEE Real-Time and Embedded Technology and Applications Symposium (RTAS)* (IEEE, 2016), pp. 1–12

[OB98]   O. Dong-Ik, T.P. Bakker, Utilization bounds for n-processor rate monotone scheduling with static processor assignment. Real-Time Syst. **15**(2), 183–192 (1998)

[RAZ17]   A. Roy, H. Aydin, D. Zhu, Energy-aware standby-sparing on heterogeneous multi-core systems, in *2017 54th ACM/EDAC/IEEE Design Automation Conference (DAC)* (IEEE, 2017), pp. 1–6

[RAZ19]   A. Roy, H. Aydin, D. Zhu, Energy-efficient fault tolerance for real-time tasks with precedence constraints on heterogeneous multicore systems, in *2019 Tenth International Green and Sustainable Computing Conference (IGSC)* (IEEE, 2019), pp. 1–8

[SK10]   G.L. Stavrinides, H.D. Karatza, Scheduling multiple task graphs with end-to-end deadlines in distributed real-time systems utilizing imprecise computations. J. Syst. Softw. **83**(6), 1004–1014 (2010)

[WBL+14]   G. Wassi, M. El Amine Benkhelifa, G. Lawday, F. Verdier, S. Garcia, Multi-shape tasks scheduling for online multitasking on fpgas, in *Reconfigurable and Communication-Centric Systems-on-Chip (ReCoSoC), 2014 9th International Symposium on* (IEEE, 2014), pp. 1–7

[WL12]   G. Wassi-Leupi, Online scheduling for real-time multitasking on reconfigurable hardware devices (2012)

[WP03]   H. Walder, M. Platzner, Online scheduling for block-partitioned reconfigurable devices, in *Proceedings of the conference on Design, Automation and Test in Europe* vol. 1 (IEEE Computer Society, 2003), p. 10290

[WZC+17]   T. Wei, J. Zhou, K. Cao, P. Cong, M. Chen, G. Zhang, X.S. Hu, J. Yan, Cost-constrained qos optimization for approximate computation real-time tasks in heterogeneous mpsocs. IEEE Trans. Comput.-Aided Des. Integr. Circ. Syst. **37**(9), 1733–1746 (2017)

[XP14]   Y. Xiang, S. Pasricha, Fault-aware application scheduling in low-power embedded systems with energy harvesting, in *Proceedings of the 2014 International Conference on Hardware/Software Codesign and System Synthesis* (ACM, 2014), p. 32

[YVH08]   H. Yu, B. Veeravalli, Y. Ha, Dynamic scheduling of imprecise-computation tasks in maximizing qos under energy constraints for embedded systems, in *Proceedings of the 2008 ASP-DAC* (IEEE Computer Society Press, 2008), pp. 452–455

[ZAZ12]   B. Zhao, H. Aydin, D. Zhu, Energy management under general task-level reliability constraints, in *2012 IEEE 18th Real Time and Embedded Technology and Applications Symposium* (IEEE, 2012), pp. 285–294

[ZCC+17]   J. Zhou, K. Cao, P. Cong, T. Wei, M. Chen, G. Zhang, J. Yan, Y. Ma, Reliability and temperature constrained task scheduling for makespan minimization on heterogeneous multi-core platforms. J. Syst. Softw. **133**, 1–16 (2017)

[ZYW+17]   J. Zhou, J. Yan, T. Wei, M. Chen, X.S. Hu, Energy-adaptive scheduling of imprecise computation tasks for qos optimization in real-time mpsoc systems, in *Proceedings of the conference on design, automation & test in Europe* (European Design and Automation Association, 2017), pp. 1406–1411

Chapter 3
Scheduling Algorithms
for Reconfigurable Systems

3.1 Introduction

Many of today's modern embedded systems is employing FPGAs within their architectures. The examples of such systems are: PDAs (personal digital assistants), mobile phones, automotive systems etc. [HKM+14]. FPGAs have also been used in computer vision, object tracking [JLJ+13], cryptography [BGHD13], and audio/video [TKG10] applications. Many of such applications are real-time in nature. For a complex safety critical system, FPGAs can act as an efficient backup platform for real-time task execution. In a typical situation, when the primary processor(s) fails, the FPGA can be used for processing the real-time tasks that were executing on the faulty primary processor(s).

However, execution of a set of hard real time tasks on FPGAs impose many challenges. First of all, it requires a well defined resource allocation and admission control mechanism. Moreover, such mechanism must guarantee timing constraints as well as high resource utilization by effectively placing hardware tasks on the 2D reconfigurable floor area.

If we recall, FPGAs consists of a 2-dimensional array of $W \times H$ as Configurable Logic Blocks (CLBs). A hardware task T_i is represented by a digital circuit which is rectangular in shape. Each task T_i consumes a sub-region $w_i \times h_i$ on the floor (having total area $W \times H$) of the FPGA. Given a set of tasks $T = \{T_1, T_2, \ldots, T_n\}$ to be executed, the placement problem can be defined as to find a subregion of size $w_i \times h_i$ for each task T_i such that no subregion overlaps with the FPGA boundaries or with other subregions (placed tasks). After placing a set of tasks, any vacant region which area is not enough to accommodate any new task, could be considered to be wasted due to fragmentation.

To minimize fragmentation, In [BKS00], Bazargan et al. proposed KAMER (Keeping All MERs), a mechanism which proceeds by maintaining a list of Maxi-

The original version of this chapter was revised: The author's name in Ref. [MVGR13] have been corrected. The corrections to this chapter are available at https://doi.org/10.1007/978-3-030-79701-0_8

K. Guha et al., *Self Aware Security for Real Time Task Schedules in Reconfigurable Hardware Platforms*, https://doi.org/10.1007/978-3-030-79701-0_3

mal Empty Rectangles (MERs). Upon arrival of a task, the algorithm places it in the bottom-left corner of the largest available MER in a worst-fit manner. After placement, the remaining vacant area of the region is partitioned using either a vertical or horizontal split to produce two empty rectangular sub-regions. Although this method produces good placement quality, a drawback is that a wrong splitting decision could cause the rejection of an otherwise feasible task. Walder et al. [WSP03] proposed an enhanced version of the Bazargan partitioner which postpones the vertical/horizontal splitting decision until the arrival of a new task in order to overcome the possibility of a wrong decision. The authors in [EFZK08] used First Fit and Best Fit placement strategies and claimed to achieve lower fragmentation and task rejection rates compared to KAMER [BKS00] and Enhanced Bazargon [WSP03] methods. However, this work maintains the occupancy status of each CLB of an FPGA in a 1D array and therefore, is susceptible to high computational overheads. A few other placement approaches aimed towards the efficient management of empty regions include algorithms by Handa et al. [HV04] and Tabero et al. [TSMM04] (which employs a Vertex List Set (VLS), where a given free space fragment is represented by a list of vertices). Ahmadinia et al. [ABBT04] proposed management of occupied area instead of free area as they observed that records for occupied spaces grow at a much slower rate than those for free spaces, making data management for the accommodation of new tasks simpler.

The above approaches are primarily oriented towards spatial management of the reconfigurable resource. Diessel and Elgindy [DE01], combined placement along with a temporal scheduling mechanism for non real-time preemptible task sets. Tasks are allocated from a ready queue starting from the bottom-left position using a first-fit strategy. When the allocator fails to accommodate the next pending task, it attempts to create additional space by preempting and locally repacking the currently running tasks utilizing spare logic resources created due to the partial execution of these tasks.

Spatio-temporal scheduling of tasks with pre-assigned priorities have also been considered for reconfigurable systems, where priority could be based on the temporal (deadline, laxity) or geometrical (size, aspect-ratio) properties of the tasks (or both) [WL12]. Walder and Platzner [WP03] used non-preemptive scheduling schemes like *First Come First Serve (FCFS) and Shortest Job first (SJF)* for block-partitioned 1D reconfigurable devices. In [CH05], authors present a scheduling mechanism called *Classified Stuffing (CS)*, where both geometrical and temporal parameters have been used to obtain priorities. The tasks are first inserted into a ready queue based on their temporal priorities (For non real-time tasks, these priorities are obtained using the *Shortest Remaining Processing Time (SRPT)* strategy, while for real-time tasks, the *Least Laxity First (LLF)* strategy is followed). The tasks in the ordered ready queue are then placed on the basis of a spatio-temporal parameter called *Space Utilization Rate (SUR)*, which is defined as the ratio between the area requirement and the execution requirement of a task. Using a 1D area model, leftmost available columns of the FPGA were filled with tasks with high SUR ($SUR > 1$), while the rightmost available columns were filled with low SUR tasks ($SUR < 1$).

Along with placement, in order to achieve high resource utilisation, the scheduling should be preemptive in nature. However, there has been very little work in this regard for FPGAs. The main reason for this is the challenges involved in storing the status of a partially completed task and restoring saved states for re-execution. Along with CLBs, the FPGA floor additionally contains Hard embedded Blocks (HBs). These HBs may include pre-fabricated blocks of BRAMs, MULs, DSPs etc. [FPMM11].

When a context switching occurs, the contexts of the tasks are stored in Flip-flops and LUT-RAMs of CLBs and other HBs. This context storing is then followed by context restore event which attempts to restore task's context as it was before the preemption, so that the execution can be resumed. Authors in [JTW07, LWH02] employed *Scan Path* generation strategy to allow task context extraction/insertion on FPGAs. However, for this preemption technique, the methodology requires task specific components known as "scan-chains" to be added with each hardware task, thus it consumes huge spatial overheads. An improvement over this technique is *bit-stream read back* methodology [JTHT10]. In [JTHT10], the authors collect FPGA configuration information by reading the bitstream through ICAP (Inter Configuration Access Port) in order to realize the context switching. Hardware task preemption on heterogeneous FPGAs is also feasible and recently demonstrated in [HTK15]. The work is similar to [JTHT10, MVGR13] and additionally allow context capture and restore of BRAMs.

3.2 Challenge for Devising Real-Time Scheduling Algorithm for FPGAs

To achieve the higher resource utilization and minimize task rejections, a different dynamic scheduling and placement approach has been discussed in [Mar14]. In this article, the author assumes that a hardware task could have multiple hardware variants (varying size) such that, larger the size of a task, faster is its performance. Selection of appropriate hardware task variants is a trade-off between maximal utilization of reconfigurable resources versus the timing requirements of the tasks. A similar algorithm which considers multiple hardware task shapes has been proposed in [WBL+14]. The authors showed that resource utilizations using conventional scheduling algorithms like EDF and LLF may be significantly improved by using the flexibility of multiple shapes, against a rigid task scenario. This work neglected reconfiguration overheads and assumed tasks to be soft real-time in nature. However, reconfiguration overhead is a major constraint which may adversely affect the temporal performance of tasks if not handled appropriately. In [LMBG09], authors described the *reuse and partial reuse* approach which allows a single configured task on the reconfigurable floor to be shared and reused among multiple applications, thus reducing the number of reconfigurations required. An important observation here is that most of the above works are primarily focused towards partially reconfigurable platforms with very few articles attempting to handle the problem for fully reconfigurable FPGAs.

As the first attempt, Danne and Platzner [DP05] has considered the scheduling preemptive real-time tasks on FPGAs. They proposed EDF-Next Fit (EDF-NF), a variant of the EDF [Liu00]. Similar approaches for FPGAs may be found in [6], [SRBS10,

BGSH12]. Recently, there has been a few contemporary multiprocessor scheduling strategies like ERfair [AS00], DP-Fair [LFS+10] etc. which provides higher resource utilisation for multiprocessor systems. The concept of such "fair" scheduling strategies is that given a set of n tasks $\{T_1, T_2, ..., T_n\}$, where T_i has an execution requirement of e_i time units, which has to be completed within a period/deadline of p_i time units. ERfair algorithm calculates *weight* of T_i as $wt_i = \frac{e_i}{p_i}$ and maintains a rate of execution for each task T_i within a scheduling time slot. Thus, ERfair suffers from huge migration and context-switching overheads. On the other hand, DP-Fair [LFS+10] enforces a maintains such rate at task period / deadline boundaries i.e. all tasks have to complete a some share of (proportional to its weight) at the end of the deadline. DP-Fair divides time into some small windows or slots, based on the task deadlines. Within that window, each task is allocated a workload. However, this constraint can only be achieved if $\sum_{i=1}^{n} e_i/p_i \leq m$, where m is the number of processor. But, these fair scheduling strategies cannot be directly employed for FPGAs due to their architectural constraints and reconfiguration overheads.

The main architectural constraints is the placement of tasks. In this chapter, we have devised real-time scheduling algorithms for FPGAs based on 2D partitioned area model where both the width W and height H of the FPGA is partitioned so that a fixed number equal sized rectangular tiles can be obtained. As there exists no partition inside the FPGA, these tiles can be called as Virtual Partitions or VPs. Now onwards, we will call each tile of the FPGA as VP. Each VP must be large enough to accommodate any task. The main reason for choosing this area model is that it may be noted that FPGAs are primarily used in embedded systems where all tasks that may possibly arrive for execution is known offine, although their actual arrival instance at runtime may not be known. Hence, the use of equi-sized tiles (2D slotted area model) with sufficient resources to execute any arbitrary task, is a legitimate and realistic assumption for embedded system.

As we have seen in our previous chapter that relaxations on the 2D slotted area model leads to flexible 2D area model which enables tasks to be placed at any arbitrary region of the FPGA. But this strategy becomes a complex placement problem which costs higher computational complexity. We can further argue that in general, relaxations on the 2D slotted area model, lead to two important drawbacks:

- Even though they improve average performance for a set of tasks, they often degrade scheduling predictability, especially in real-time systems. This is because, it becomes more complex in this case to deterministically account for the spatio-temporal capacity available in the worst-case, within a given time interval in future.
- These models tend to incur much higher overheads at each spatial scheduling point, making the complexity of the overall spatio-temporal scheduling problem very expensive towards online application.

Majority of the existing scheduling algorithms for FPGAs mainly focus on the problem of efficient task placement on the FPGA floor such that fragmentation is minimized. Comparatively a very few research has been conducted towards scheduling so that the deadlines of real-time tasks are effectively managed. Moreover, research towards real-time scheduling have mostly been confined to the handling of soft real-time tasks and aimed for partially reconfigurable FPGA. In comparison, there is very little researches that address the hard real-time scheduling and effective resource management in fully reconfigurable systems.

In this chapter, we will devise our scheduling strategies for fully and partially reconfigurable FPGAs by assuming a 2-dimensional FPGA and the FPGA floor is equi-partitioned into a set of VPs. It has been assumed that each VP ha sufficient resources to accommodate a task. We consider fully reconfigurable FPGAs with reconfiguration overhead denoted as O_{frg}. However, it has to be noted that for fully reconfigurable FPGA, all VPs have to be reconfigured simultaneously. In partially reconfigurable FPGA, the reconfiguration overhead is denoted as O_{prg}. For partially reconfigurable FPGA, we can reconfigure a VP while the other VPs can operate seamlessly. Now, it may be noted that the configuration time of an FPGA depends upon two factors, one is the initialization time which requires only one time, on power up of the system and another is bit-stream loading time, to configure the FPGA with specific hardware circuit. Here, we are not considering the initialization time, the O_{frg} & O_{prg} are the configuration time or more specifically bit-stream loading time. Now configuration time depends on the bit-stream size, the clock frequency and the bandwidth of configuration port (Described later Section).

As the modern FPGAs support high data width, parallel configuration port and high (50 MHz) configuration clock frequency, O_{frg} differs due to variable bit-stream size as the information needed to fully configure the floor of the small FPGA varies from larger one [Sta11]. In case of O_{prg}, the partial reconfiguration time is directly proportional to the size of the reconfigurable region or more specifically to the bit-stream size needed to partially reconfigure the region [HSH09]. A study has been made to have a view on the different full configuration times for Xilinx FPGAs and found that it varies from around 3 ms to 30 ms depending upon the available floor size. The real-time scheduling strategies for FPGAs must incorporate these recon-figuration overhead (O_{frg} and O_{prg}) for fully and partially reconfigurable systems respectively). It has to be ensured that by considering these overheads, the timing constraints of the system do not get violated.

This chapter is organized as follows. The next section provides a brief discussion on the system model and assumptions. Detailed descriptions of our proposed scheduling strategies (for fully reconfigurable system and partially reconfigurable systems), have been presented along with illustrative examples in Sect. 3.4. Section 3.5 describes how the dynamic tasks can be handled by both the strategies. Section 3.6 represents experimental results along with analysis and discussion on the same. The chapter finally concludes in Sect. 3.7.

3.3 System Model and Assumptions

The overall system model contains FPGA, a General Purpose Processor (GPP) and memory, as shown in Fig. 3.1. The bitstream of a task (bit file or configuration file which is the executable unit of a hardware task) is stored in the memory unit. GPP

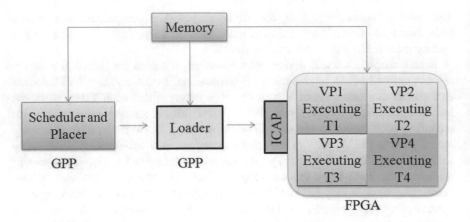

Fig. 3.1 System model

is responsible for executing hardware tasks on the different tiles/VPs (Virtual Partitions) of FPGAs by loading bitstream file from the memory unit to the configuration memory of the FPGA via ICAP (Internal Configuration Access Port). The architecture of the FPGA is similar to that of the Xilinx Virtex series of FPGAs. The FPGA consists of CLBs and other Hard Blocks (HBs) like Block RAMs (BRAM), Multipliers (MULs), I/O Blocks (IOBs), Clocks (CLKs) etc.

The floor of the FPGA is equi-partitioned into M VPs. Each VP can execute any hardware task irrespective of their resource requirements. The considered task set are preemptive and hence, it will consume context switch overheads. This overheads are determined by reconfiguration overheads of the FPGA. For partially reconfigurable FPGA, individual VPs can be reconfigured in an asynchronous fashion but for a fully reconfigurable FPGA, context switching becomes a global event because all the VPs have to be reconfigured synchronously.

In order to reconfigure a VP in FPGA, (of area say, α), we need to halt/stop the currently running task within that VP and then the task's bitstream is extracted/captured from the configuration memory through ICAP. In [JTE+12], the authors showed that actual context of a task is present within less than 8% of its total bitstream and while conducting the context switch, only this part of the bitstream requires to be extracted. Thanks to this selective read-back mechanism, with the aid of this technique, actual extraction overhead is reduced to 1/10th of the full bitstream read-back time. Once the extraction is complete, the captured task bitstream is manipulated [JTHT10] and the new bitstream is created. For the context restoration, the updated bitstream is loaded.

The time required for (say, TR) extraction/restoration of the region α is depends on the the size of the bitstream (BS) and the configuration clock frequency ($CCLK$) of the controller and its data bus width (DBW). TR can be calculated as [Tap13]:

$$TR = BS/(CCLK \times DBW) \tag{3.1}$$

To quantify TR for Xilinx Virtex-4 FPGA which has an area of 52×128 [Xil10], full configuration bitstream of size 2.21 MB [Xil09], $CCLK = 100\,\text{MHz}$ and $DBW = 32$ bit [Xil10]. From these values, the total context restoration time becomes $\approx 6\,\text{ms}$, while the extraction time is ($\frac{1}{10} \times 6\,\text{ms}$) = 0.6 ms, which we assumed to be 1 ms. The bit manipulation time and the overhead for new bitstream formation is $\approx 5\,\text{ms}$. Hence, the total context switching overhead (O_{frg}) becomes 1 ms+ 5 ms+ 6 ms=12 ms (extraction + manipulation + restoration). (O_{prg}) for partially reconfigurable FPGA having M VPs can be calculated as: $O_{prg} = O_{frg}/M$. Thus, if Virtex-4 is partitioned into four equi-sized VPs, then $O_{prg} = \frac{12}{4} = 3$ ms.

3.4 Scheduling Strategies

Given a set of n tasks $\{T_1, T_2, \dots, T_n\}$ to be scheduled on M VPs $\{VP_1, VP_2, \dots, VP_m\}$ of an FPGA, the proposed scheduling algorithms maintain the time partitioned into slices / windows (the rth time-window is denoted by tw_r, difference between $(r-1)$th and rth task deadlines) demarcated by the task deadlines. At any time-window boundary, each task T_i has to complete a workload-quota WQ_i^r for the time-window tw_r (having length twl_r) given by: $WQ_i^r = \lceil \frac{e_i}{p_i} \times twl_r \rceil$. The sum of the workload-quota of all tasks can be calculated as: $sum\_WQ^r = \sum_{i=1}^{n} WQ_i^r$, and the total system capacity becomes ($twl_r \times M$) over the interval twl_r, Now in order to achieve a feasible schedule both for fully and partially reconfigurable systems, following conditions needs to be satisfied:

$$sum\_WQ^r \leq twl_r \times M \qquad (3.2)$$

3.4.1 Scheduling Algorithm for Full Reconfigurable Systems

Context switching in fully reconfigurable FPGAs can only be realised through synchronous configuration of VPs and each such context switch will consume a full reconfiguration overhead. The total overhead that for all VPs is given by: $O_{frg} \times M$, where O_{frg} denotes the full reconfiguration time. Now, this overhead can be compensated from the available slack (if any) at time-window duration twl_r and is given by: $twl_r \times M - sum\_WQ^r$. Hence, in order to achieve the feasible solution, the following condition needs to be hold true i.e.:

$$O_{frg} \times M \leq twl_r \times M - sum\_WQ^r \qquad (3.3)$$

The maximum number of context switches/ full reconfiguration C_r that can be afforded within twl_r is:

$$C_r = \lfloor \frac{twl_r \times M - sum\_WQ^r}{O_{frg} \times M} \rfloor \tag{3.4}$$

Therefore, the scheduling algorithm checks whether the condition in Eq. 3.3 is satisfied or not and calculates the maximum number of affordable full reconfigurations (C_r) using Eq. 3.4. These reconfigurations is required to allocate te tasks to the M VPs. The time duration between two consecutive reconfiguration events is denoted as *time-frame*. Thus, within a time-window tw_r, there will be C_r time-frames and denoted as $g_1, g_2, \ldots, g_{C_r}$. These time-frames are equi-sized and the length is denoted as G_r. It should be noted that within a time-frame in a vp, tasks will be executed.

At the beginning of each time-frame in twl_r, the scheduling algorithm selects M tasks with the *highest remaining workload-quota* and executes them for the duration of a time-frame, on the M available VPs. Hence, T_i requires $\lceil \frac{WQ_i^r}{G_r} \rceil$ number of time-frames to complete its workload-quota WQ_i^r. All tasks will finis their allotted workload-quota, if te total number of time-frames required by all the tasks, ($\sum_{i=1}^{n} \lceil \frac{WQ_i^r}{G_r} \rceil$) becomes at most the total number of available time-frames ($C_r \times M$). Thus,

$$\sum_{i=1}^{n} \lceil \frac{WQ_i^r}{G_r} \rceil \leq C_r \times m \tag{3.5}$$

As a task can not execute in parallel in multiple VPs, and $C_r \times G_r$, denotes the the maximum execution time for a task within a time-window.

$$WQ_i^r \leq C_r \times G_r \tag{3.6}$$

Example 1 Let us assume five tasks $T_1, T_2, \ldots, T_5$, having weights ($\frac{e_i}{p_i}$) 14/60, 36/90, 42/60, 72/90, 54/90 and 54/90. Without loss of generality, we represented the real-time constraint as: "T_i *must have to compute e_i number of instructions*[1] *within p_i time units*". Also, we have $M = 4$ (4 partitions), $O_{frg} = 6$ time units[2] and the first time-window $tw_1 = 60$. The task workload-quota to be executed within the interval tw_1 are: $WQ_1 = 14$; $WQ_2 = 24$; $WQ_3 = 42$ and $WQ_4 = WQ_5 = 36$. As $sum\_WQ = 152$, $tw_1 \times m = 240$ and $OF_{rg} \times m = 24$. $C_r = \lfloor (240 - 152)/24 \rfloor = 3$ (refer Eq. 3.4). Hence, we may allow three time-frames in the time-window, each of length 14 time units. As there are three time frames hence, each task will have three subtasks in order to complete its execution within a time slice. We have represented this subtask in each frame as T_{ij} where j denotes the subtask id and it should be same as the frame id.

The condition in Eq. 3.5 is also satisfied. After a full reconfiguration from 0 to 6 time slots, the first time-frame executes with the tasks $T_{31}, T_{41}, T_{51}, T_{21}$ within 6 to

---

[1] Assuming 1 instruction is completed in 1 time unit using standard FPGA clock frequency.

[2] Assuming a hypothetical FPGA with small floor area.

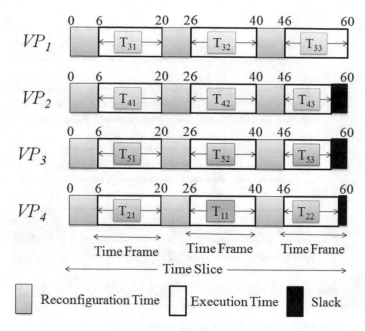

Fig. 3.2 Task scheduling for full reconfiguration system

20 time units. The 2 nd full reconfiguration occurs between 20 and 26 time units. Tasks T_{32}, T_{42}, T_{52}, T_{12} are selected and executed in the second time-frame within the time unit interval 26–40. Finally tasks T_{33}, T_{43}, T_{53}, T_{23} executed in third time-frame from interval 46 to 60. Thus, each task finishes its workload-quota within the time-window/slice as illustrated in Fig. 3.2.

3.4.2 Scheduling Algorithm for Partially Reconfigurable Systems

The main architectural advantage of a partially reconfigurable FPGA is that, unlike full reconfigurable FPGA, context switching (reconfiguration) is not a global event and is localized to individual VPs. In addition, O_{prg} is lower than O_{frg}.

If the condition in Eq. 3.2 is satisfied, the scheduling strategy will partition the task set into M disjoint subsets as follows: the tasks are allocated from the first VP, VP_1. While allocating tasks within a VP, it has to be ensured that the sum of task workload-quota along with O_{prg} should be less than length of the time-window twl_r. It should be noted that all VPs consumes O_{frg} at the beginning of the time-window; hence, the remaining capacity rc_i for each VP V_i is given by: $rc_i = twl_r - O_{frg}$. A task say T_j, is allocated to a VP V_i by consuming O_{prg} and hence, rc_i of V_i becomes: $rc_i = rc_i - WQ_j^r - O_{prg}$. However, Such an allocation is only allowed if

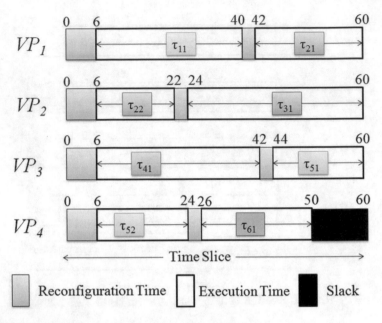

Fig. 3.3 Task scheduling for partially reocnfigurable FPGA

$rc_i - WQ_j^r \geq 0$. Otherwise, T_j must be divided into two subtasks; the first task T_{j1} with a workload-quota value $WQ_{j1}^r = rc_i$, is executed in V_i while the second part with $WQ_{j2}^r = WQ_j^r - rc_i$, is executed in the next VP V_{i+1} provided $i < M$. If V_i be the last available VP ($i = M$), then scheduling will not commence in time-window tw_r as the system capacity is not sufficient.

Example 2 Let us consider the partially reconfiguarble FPGA contains 4 VPs and the same set of 6 tasks as used in our previous example however, T_4's execution requirement e_4 is increased from 72 to 74. Hence, the workload-quota WQ_4^1 within the first time-window ($tw_1 = 60$ ms) becomes 49. We assumed that $O_{frg} = 6$ ms and $O_{prg} = 2$ ms. The sum of workload-quota (sum_WQ^1) is 193.

This task set can't be scheduled on fully reconfigurable FPGA. The initial capacity of each of the four VPs will be $twl_1 - O_{frg} = 54$ ms. Figure 3.3 shows the schedule generated by the scheduling algorithm for partially reconfigurable FPGA. It may be noted that while T_1, T_2 and T_5 is executed completely on a single VP, T_3, T_4 and T_6 is executed partially on two different VPs with their subtasks.

3.4.2.1 Avoidance of ICAP Conflict

Modern FPGAs contain a single configuration port i.e. ICAP and hence, no two VPs can be reconfigured simultaneously. This essentially means that, if we load a task bitstream in one VP via ICAP then at the same time, we cannot load any other task's bitstream as the ICAP port is busy. Our scheduling algorithm handles the situation in following way. Scheduling algorithm proceeds VP by VP, starting from the first VP. When it generates the schedule for the first VP, the time instants for reconfigurations are also obtained. Thus, scheduler has these reconfiguration instants when it generates schedule for the second VP. So, if any reconfiguration instant of the second VP overlaps with the reconfiguration instant of the first VP, then the execution of the task workload-quota is adjusted so that any overlapping can be avoided. Similarly, for the third VP, the task schedules along with reconfiguration instants for the first and second VPs are already known. Therefore, if any reconfiguration event of the third VP overlaps with those of other VPs, then the task workload-quota for the third VP is adjusted. This procedure will continue till the last VP is reached.

3.4.3   Handling Dynamic Tasks

In dynamic real-time system, task arrival and departure is a dynamic process. However, it as to be ensure that the newly arrived task will only be accepted if the inclusion of a new task will not violate the deadlines of already accepted tasks. We will now discuss how dynamic tasks can be handled in case of fully and partially reconfigurable FPGAs, respectively.

3.4.4   For Fully Reconfigurable FPGAs

Let us assume task T_i leaves the system, as a result, future time-frames dedicated for T_i's execution in the present time-window becomes free. These free time-frames can be used for the allocation of newly arrived tasks.

Let us assume task T_i arrives at a time instant say t_u in the middle of time-frame (corresponding time-window starts at time t_v), T_i has to wait till the current time-frame completes, as reconfiguration is not allowed between a time-frame. T_i may be allocated to a VP for execution in time-frames, where there exists free unallocated VPs. These time-frames will be executed at higher priority so that T_i meets the deadline in case, $d_i < (t_v + twl_r - t_u)$, where d_i is the deadline.

At the end of this allocation, the amount of execution could be be completed by the tw_r is calculated. If it is found that T_i's deadline can be met then it's remaining execution time and period will be updated.

3.4.5   For Runtime Partially Reconfigurable Systems

In case of partially reconfigurable FPGA, at each time-window task T_i is partitioned into subtasks and allocated specific sub-interval(s) on one or two VPs. For example,

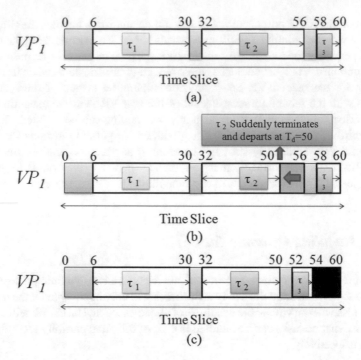

Fig. 3.4 Handling dynamic task departure

from Fig. 3.3, it may be observed that task T_3 has been partitioned and the subtasks are allocated on V_2 and V_1, respectively.

If a task T_i departs from the system within a time-window, then the dedicated VP becomes free and can be utilised for allocating a dynamically arriving task within the time-window. However, in order to ensure that no VP remains unutilised due to certain termination of a task thus, the entire schedule (task execution sequence) after T_i's completion is preponed and executed starting from t_d, departure instance.

Figure 3.4.a shows a typical schedule in a time-window of length 60 on a VP, V_1. Let us assume that task T_2 departs at time $t_d = 50$ (This is shown in Fig. 3.4b). Then the rest of the schedule is preponed and executed from time 50, as shown in Fig. 3.4c.

Let us consider T_i dynamically arrives at time (say t_a) in the middle of the rth time-window tw_r, its workload-quota (WQ_i) for the remaining part of the time-window is calculated and allocated in the free sub-intervals (if any) of one or more VPs starting with the VP having the highest remaining sub-interval. If the VP $(V_k$ say) with the highest remaining sub-interval is in idle state at time t_a (which means that V_k currently has no task to execute for the rest of the time slice), then the entire workload-quota of T_i may be accommodated in V_k. Otherwise, the WQ_i of T_i must be partitioned and allocated to more than one time-window. While making such an allotment of free sub-intervals for T_i at time t_a, T_i is assigned the highest priority (this is done by postponing the rest of the schedules on these VPs to the end of the time-window)

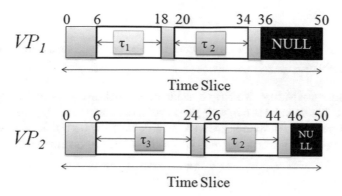

Fig. 3.5 Schedule in tile V_1 & V_2

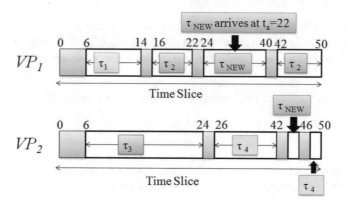

Fig. 3.6 Handling dynamic task arrival

to account for the case in which T_i's deadline is earlier than the next time-window boundary. This task allocation mechanism has been depicted in Fig. 3.5.

Figure 3.5 shows the schedules for an arbitrary time-window on two VPs V_1 and V_2. At a time $t_a = 22$ when a new task T_{new} arrives with workload-quota $WQ_i = 18$, then the task allocation scheme is described in Fig. 3.6.

3.5 Experiments and Results

We have evaluated the scheduling algorithms through simulation experiments followed by hardware implementations. *Task Rejection Rate (TRR)* has been used as the principal metric for the evaluation. *TRR* can be defined as the ratio between the total number of tasks rejected and total number of tasks arrived. That is,

$$TRR = (\nu/\psi) \times 100 \qquad\qquad (3.7)$$

where, ν and ψ denote the total number of rejected tasks and total number of arrived tasks, respectively.

Experimental Setup: We have considered normal distributions for the task set generation.. Tasks weights ($wt_i = \frac{e_i}{p_i}$) and task execution periods (p_i) have also been taken from normal distributions. Given the task weights, we obtain WL, work-load by summing up the the task weights. Utilization U can be obtained as:

$$U = \frac{WL}{M} \times 100 \ (\%) \qquad\qquad (3.8)$$

where, M denotes the number of VPs.

Now, we will introduce various parameters for our simulation.

1. *Number of VPs M*: The FPGA contains into 2, 4, 6 and 8 VPs.
2. *Utilization U*: Task utilization values varying from $U = 50\%$ to $U = 90\%$.
3. *Average individual tasks weight μ_{wt}*: μ_{wt} have been considered in the range 0.1–0.5.

For a given the system utilisation (U), the average number of tasks (ρ) can be derived as:

$$\rho = \frac{U \times M}{100 \times \mu_{wt}} \qquad\qquad (3.9)$$

The total schedule length is 200,000 time-slots. All results are generated by running 40 different instances and taking average of these 40 runs.

3.5.1 Results and Analysis

Fig. 3.7 shows the *TRR* suffered by the fully reconfigurable FPGA having eight VPs ($M = 8$) when the utilisation varies from 50 to 90%. From the figure, it can be observed that as the utilisation increases the *TRR* also increases. The main reason for this is that higher utilisation causes higher number of tasks (ref Eq. 3.9) in the system. As the number of tasks increases, the slack inside the VPs decreases and thus, the less number of tasks are scheduled. Specifically, higher U, the LHS of Eq. 3.5 becomes higher. Thus, the probability of failure of Eq. 3.5 increases. Another interesting observation is that as the μ_{wt} increases from 0.1 to 0.5, *TRR* decrease. The main reason behind this is that, for a fixed utilisation, higher μ_{wt} refers less number of tasks (refer Eq. 3.9) and thus, rejection rate decreases.

Figure 3.8 shows the variation of the *TRR* with respect to number of VPs, M. It can be observed that as the VPs increase while maintaining the U fixed at 70%, the *TRR* decreases. This could be attributed to the fact that as the VPs increase, the system wide capacity also increase and thus, more number of tasks can be accommodated

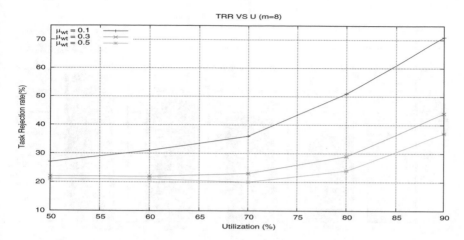

Fig. 3.7 *TRR* versus U; $M = 8$

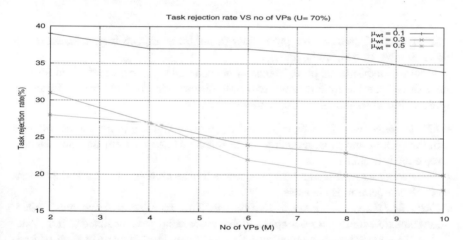

Fig. 3.8 *TRR* versus M; $U = 70\%$

Table 3.1 TRR versus O_{frg}; $\mu_{wt} = 0.3$; $M = 8$ and $U = 70\%$

O_{frg} (ms)	12	18	24	30
TRR (%)	23	32	42	55

in the system and thus, *TRR* decreases Similarly, as the μ_{wt} increases the number of task in the system decreases and this attributes to lower rejections.

Table 3.1 depicts how the *TRR* varies when the full reconfiguration overhead O_{frg} varies. Here, we fixed the VP at 8, utilisation U was fixed at 70%, and μ_{wt} is fixed at 0.3. Now, it can be observed that higher value of O_{frg} causes higher value of *TRR*. This is mainly due to the fact that when the O_{frg} increases C_r decreases (refer,

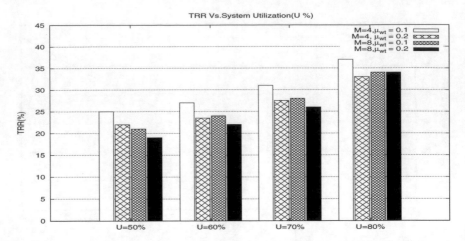

Fig. 3.9 *TRR* versus *U*

Eq. 3.4) and as C_r reduces, the probability of failure of Eq. 3.5 increases, causing a higher number of tasks to be rejected.

Figure 3.9 exhibits the performance of our scheduling algorithm devised for partially reconfigurable systems. We will now discuss the obtained trends from this figure and its corresponding justifications.

- *TRR increases with U*: This is because higher U is causing more number of tasks in the system and thus, reducing the available capacity within the tiles. Hence, rejection is higher.
- *TRR decreases with μ_{wt}*: Higher μ_{wt} refers to less number of tasks in the system and thus, rejection is also less.
- *TRR is less than fully reconfigurable systems*: This mainly due to two facts. Firstly, the context switching is no longer a global event rather it is localised to individual VPs. This asynchronous reconfiguration leverages the better utilisation of individual VPs and thus, rejection is less. Another reason is that the reconfiguration overhead for partially reconfiguarable FPGAs is less than the fully reconfigurable systems.
- *TRR decreases with the increase in M*: This observation can be attributed to the fact that, higher be the number of M lower be the O_{prg} (refer, Sect. 3.3). As the O_{prg} decreases rejection rate also decreases.

Table 3.2 shows *TRR* suffered for a partially reconfigurable FPGAs. From this Table, three trends can be observed. First, as the utilisation increases, *TRR* also increases, this is mainly because, higher U is causing large number of tasks in the system and as a result the slack within the VPs are reducing causing higher rejections. Similarly, as μ_{wt} increases *TRR* decreases. But, it has to be noted that for fixed parameters (i.e., M, U and μ_{wt}) partially reconfigurable systems suffer less rejections than fully reconfigurable systems this mainly for two reasons i. the partial

Table 3.2 Task rejection rate (%)

U (%)	M = 2			M = 4			M = 8		
	$\mu_{wt} =$ 0.1	$\mu_{wt} =$ 0.3	$\mu_{wt} =$ 0.5	$\mu_{wt} =$ 0.1	$\mu_{wt} =$ 0.3	$\mu_{wt} =$ 0.5	$\mu_{wt} =$ 0.1	$\mu_{wt} =$ 0.3	$\mu_{wt} =$ 0.5
50	30	28	26	25	21	19	21	18	16
60	31	30	28	27	23	22	24	21	20
70	35	33	32	31	27	26	28	25	22
80	42	40	38	37	32	30	34	33	31
90	45	43	425	42	37	335	39	37	35

reconfiguration overhead is lower than the full reconfiguration overhead. ii. Each individual VPs can be reconfigured without interrupting other VPs, this asynchronous reconfiguration is another factor for less rejections. Another interesting observation is that *TRR* decreases as the number of VPs increase. This observation can be supported by the fact that the ($O_{prg} = \frac{O_{frg}}{M}$); as shown in Sect. 3.3) O_{prg} decreases with increase in the number of VPs.

3.6 Hardware Prototype for Multiple Tasks Processing on FPGA

A simple prototype system for hardware task multitasking has been implemented on ZYNQ:ZC702 FPGA board based on our scheduling approach with benchmark task sets. *We would like to emphasize that with this simple HW prototype, we want to develop a proof of concept system that multitasking on FPGAs with "tiled" architecture is indeed feasible.* Our hardware prototype works as follows:

- The the task scheduling algorithm runs on ARM processor, located in the ZYNQ board.
- We have taken three applications from *EPFL Combinational Benchmark Suite* [AGDM15] i.e. *Context-adaptive variable-length coding (CAVLC), Decoder (Dec), Integer to float conversion (I2F).* The actual execution cost of sample tasks is shown in Table 3.3.
- Each of the application/task is coded in VHDL and performance of each such task is measured through proper test-bench.
- At the beginning, hardware tasks are stored in its executable format (as *.bit*) in an external memory.
- Three RRs (Reconfigurable Regions) are created and three tasks were mapped into these RRs.
- Tasks were executed for length of pre-calculated *time-frame* (as determined by the s) till a full reconfiguration overhead.

Table 3.3 Benchmark tasks execution overhead on ZYNQ

Tile	Tasks	Execution overhead (ticks)
RR1	CAVLC	465335
RR2	Dec	211654
RR3	I2F	106723

Table 3.4 TRR versus U

U %	30	40	50	60
TRR (%)	10	16	22	28

The ZYNQ FPGA contains two types of PEs that is the FPGA fabric and dual core ARM Cortex-A9 processor. The portion which contains ARMs are termed as PS (Processing System) and the region with FPGA logic is termed as PL or Programmable logic region [CEES14]. In PL, we have created VPs (using Xilinx Vivado tool) so that the hardware tasks can be executed. In PS, ARMs are prefabricated and available in IC format.

Customization of the platform: For the evaluation of our scheduling strategies, we have customized the architecture of ZYNQ platform.

- In the PL, the VPs are created and the execution of hardware task on that VP is designed by writing the placement constraints on the UCF [CEES14] file. Here, the UCF stands for user constraint file.
- These VPs are operating with clock (FCCLK) of 50 MHz frequency.
- One of the available ARM core is used for the execution of our scheduling algorithm and the frequency of ARM core is 650 MHz.
- As the algorithm is executed in the PS side and hardware tasks are executed in the PL side. Hence, PS and PL needs to be tightly coupled. PS and PL communications are ensured through GP0 and GP1 port. PS and PL are connected with On-Chip Memory using AXI interface.

Table 3.4 shows the TRR obtained by varying U, where the number of VPs are fixed at four. From the obtained results, it can be concluded that the hardware results follow the similar trends as obtained in software simulations.

3.7 Conclusion

Scheduling of real-time tasks on FPGAs is a challenging problem. One has to ensure that the proposed algorithm should achieve high resource utilization. Challenges are mainly due to two facts. Firstly, effectively placing hardware tasks on FPGAs such that the FPGA remains less fragmented and secondly, the reconfiguration overheads require to be judiciously handled so that deadlines are not jeopardized.

In this chapter, we have shown two scheduling methodologies for real-time task sets on fully and partially reconfigurable FPGAs. We have followed 2D slotted area model. The dynamic task handling mechanisms have also been discussed. Simulation results reveal that the proposed strategies are able to achieve higher utilisation with lower task rejection rate under various simulation scenarios. The software outcomes are also validated through real hardware implementation.

References

[ABBT04] A. Ahmadinia, C. Bobda, M. Bednara, J. Teich, A new approach for on-line placement on reconfigurable devices, in *18th International on Parallel and Distributed Processing Symposium, Proceedings* (IEEE, 2004), p. 134

[AGDM15] L. Amarú, P.-E. Gaillardon, G. De Micheli, The EPFL combinational benchmark suite, in *Proceedings of the 24th International Workshop on Logic & Synthesis (IWLS)*, number EPFL-CONF-207551 (2015)

[AS00] J.H. Anderson, A. Srinivasan, Early-release fair scheduling. ECRTS **2000**, 35–43 (2000)

[BGHD13] S. Bhasin, S. Guilley, A. Heuser, J.-L. Danger, From cryptography to hardware: analyzing and protecting embedded Xilinx Bram for cryptographic applications. J. Crypt. Eng. **3**(4), 213–225 (2013)

[BGSH12] L. Bauer, A. Grudnitsky, M. Shafique, J. Henkel, Pats: a performance aware task scheduler for runtime reconfigurable processors, in *2012 IEEE 20th Annual International Symposium on Field-Programmable Custom Computing Machines (FCCM)* (2012), pp. 208–215

[BKS00] K. Bazargan, R. Kastner, M. Sarrafzadeh, Fast template placement for reconfigurable computing systems. IEEE Des. Test Comput. **17**(1), 68–83 (2000)

[CEES14] L.H. Crockett, R.A. Elliot, M.A. Enderwitz, R.W. Stewart, *The Zynq Book: Embedded Processing with the Arm Cortex-A9 on the Xilinx Zynq-7000 All Programmable Soc* (Strathclyde Academic Media, 2014)

[CH05] Y.-H. Chen, P.-A. Hsiung, Hardware task scheduling and placement in operating systems for dynamically reconfigurable SOC. Embed. Ubiquitous Comput.-EUC **2005**, 489–498 (2005)

[DE01] O. Diessel, H. Elgindy, On dynamic task scheduling for FPGA-based systems. Int. J. Found. Comput. Sci. **12**(05), 645–669 (2001)

[DP05] K. Danne, M. Platzner, Periodic real-time scheduling for FPGA computers, in *Third International Workshop on Intelligent Solutions in Embedded Systems* (2005), pp. 117–127

[EFZK08] M. Esmaeildoust, M. Fazlali, A. Zakerolhosseini, M. Karimi, Fragmentation aware placement algorithm for a reconfigurable system, in *Second International Conference on Electrical Engineering, 2008. ICEE 2008* (IEEE, 2008), pp. 1–5

[FPMM11] U. Farooq, H. Parvez, H. Mehrez, Z. Marrakchi, Exploration of heterogeneous FPGA architectures. Int. J. Reconfig. Comput. **2011**, 2 (2011)

[Gua+07] N. Guan et al., Improved schedulability analysis of edf scheduling on reconfigurable hardware devices, in *2007 IEEE International Parallel and Distributed Processing Symposium* (2007)

[HKM+14] T. Hayashi, A. Kojima, T. Miyazaki, N. Oda, K. Wakita, T. Furusawa, Application of FPGA to nuclear power plant i&c systems, in *Progress of Nuclear Safety for Symbiosis and Sustainability* (Springer, 2014), pp. 41–47

[HSH09]   P.-A. Hsiung, M.D. Santambrogio, C.-H. Huang, *Reconfigurable System Design and Verification*, 1st edn. (CRC Press Inc, Boca Raton, FL, USA, 2009)

[HTK15]   M. Happe, A. Traber, A. Keller, Preemptive hardware multitasking in reconos, in *Applied Reconfigurable Computing* (Springer, 2015), pp. 79–90

[HV04]   M. Handa, R. Vemuri, An efficient algorithm for finding empty space for online FPGA placement, in *Proceedings of the 41st annual Design Automation Conference* (ACM, 2004), pp. 960–965

[JLJ+13]   J. Jin, S. Lee, B. Jeon, T.T. Nguyen, J.W. Jeon, Real-time multiple object centroid tracking for gesture recognition based on FPGA, in *Proceedings of the 7th International Conference on Ubiquitous Information Management and Communication* (ACM, 2013), p. 80

[JTE+12]   K. Jozwik, H. Tomiyama, M. Edahiro, S. Honda, H. Takada, Comparison of preemption schemes for partially reconfigurable FPGAs. IEEE ESL **4**(2), 45–48 (2012)

[JTHT10]   K. Jozwik, H. Tomiyama, S. Honda, H. Takada, A novel mechanism for effective hardware task preemption in dynamically reconfigurable systems, in *2010 International Conference on Field Programmable Logic and Applications (FPL)* (IEEE, 2010), pp. 352–355

[JTW07]   S. Jovanovic, C. Tanougast, S. Weber, A hardware preemptive multitasking mechanism based on scan-path register structure for FPGA-based reconfigurable systems, in *Second NASA/ESA Conference on Adaptive Hardware and Systems, 2007. AHS 2007* (IEEE, 2007), pp. 358–364

[LFS+10]   G. Levin, S. Funk, C. Sadowski, I. Pye, S. Brandt, Dp-fair: a simple model for understanding optimal multiprocessor scheduling. ECRTS **2010**, 3–13 (2010)

[Liu00]   J.W.S. Liu, *Real-Time Systems*, 1st edn. Prentice Hall (2000)

[LMBG09]   Y. Lu, T. Marconi, K. Bertels, G. Gaydadjiev, Online task scheduling for the fpga-based partially reconfigurable systems, in *Reconfigurable Computing: Architectures, Tools and Applications* (Springer, 2009), pp. 216–230

[LWH02]   W.J. Landaker, M.J. Wirthlin, B.L. Hutchings, Multitasking hardware on the slaac1-v reconfigurable computing system, in *Field-Programmable Logic and Applications: Reconfigurable Computing Is Going Mainstream* (Springer, 2002), pp. 806–815

[Mar14]   T. Marconi, Online scheduling and placement of hardware tasks with multiple variants on dynamically reconfigurable field-programmable gate arrays. Comput. Electr. Eng. **40**(4), 1215–1237 (2014)

[MVGR13]   A. Morales-Villanueva, A. Gordon-Ross, Htr: on-chip hardware task relocation for partially reconfigurable FPGAs, in *Proceedings of the 9th International Conference on Reconfigurable Computing: Architectures, Tools, and Applications, ARC'13* (2013), pp. 185–196

[SRBS10]   M.D. Santambrogio, V. Rana, I. Beretta, D. Sciuto, Operating system runtime management of partially dynamically reconfigurable embedded systems, in *2010 8th IEEE Workshop on Embedded Systems for Real-Time Multimedia (ESTIMedia)* (2010), pp. 1–10

[Sta11]   E. Stavinov. *100 Power Tips for FPGA Designers*, 1st edn (CreateSpace, 2011)

[Tap13]   S. Tapp, Bpi fast configuration and impact flash programming with 7 series FPGAs. *XAPP-587, Xilinx Application Notes* (2013)

[TKG10]   D. Theodoropoulos, G. Kuzmanov, G. Gaydadjiev, A 3d-audio reconfigurable processor, in *Proceedings of the 18th Annual ACM/SIGDA International Symposium on FPGAs* (2010), pp. 107–110

[TSMM04]   J. Tabero, J. Septién, H. Mecha, D. Mozos, A low fragmentation heuristic for task placement in 2d rtr hw management, in *FPL* (Springer, 2004), pp. 241–250

[WBL+14]   G. Wassi, M.E. Amine Benkhelifa, G. Lawday, F. Verdier, S. Garcia, Multi-shape tasks scheduling for online multitasking on FPGAs, in *2014 9th International Symposium on Reconfigurable and Communication-Centric Systems-on-Chip (ReCoSoC)* (IEEE, 2014), pp. 1–7

[WL12]   G. Wassi-Leupi, Online scheduling for real-time multitasking on reconfigurable hardware devices (2012)

[WP03]   H. Walder, M. Platzner, Online scheduling for block-partitioned reconfigurable devices, in *Proceedings of the Conference on Design, Automation and Test in Europe-Volume 1* (IEEE Computer Society, 2003), p. 10290

[WSP03]   H. Walder, C. Steiger, M. Platzner, Fast online task placement on FPGAs: free space partitioning and 2d-hashing, in *Proceedings. International on Parallel and Distributed Processing Symposium, 2003* (IEEE, 2003), p. 8

[Xil09]   Xilinx. Virtex-4 FPGA configuration user guide, xilinx. *June* (2009)

[Xil10]   Incorporation Xilinx, Virtex-4 family overview. *Tech. Doc. DS112 (v2. 0)* (2010), pp. 1–8

Part III
Security

Chapter 4
Introduction to Hardware Security for FPGA Based Systems

4.1 Introduction

Ensuring security for computer systems is of paramount importance. Analyzing various forms of attacks and defining strategies to prevent them is essential to generate trust among users. In general, to make a system reliable, system designers need to satisfy the basic three requirements, i.e. ensure confidentiality or prevent unauthorized observing of data or information, ensure integrity or prevent unauthorized change of data and ensure availability or facilitate authorized access to information or data at any instant of time and generate proper results within time. These three are commonly known as the CIA requirements [BT18]. However, with time, new attacks have arose like power dissipation attacks that may affect the green computing factor of a system or may drain the power budget of the system and cause early expiry of the system [Guh20, GMSC20]. Hence, it is the responsibility of system designers to analyze new and potential forms of threats that may arise with time and develop security strategies to mitigate them.

Previously, hardware was considered trusted and immune to attacks. The key focus was made on attacks related to softwares [BHBN14]. Software security was the key focus among researchers, where a wide variety of software attacks related to computer systems was analyzed and several solutions were also proposed, which involved static authentication mechanisms and dynamic monitoring schemes. Direct task execution on hardware was a key used by researchers to eradicate the software threats.

In the early phases, hardware security was confined to implementation dependent vulnerabilities associated with cryptographic chips, that lead to information leakage or revealing of secret key to adversaries [LJM13, GSC15]. However, with the advent of globalization in system design, threats related to hardware became prominent. In the era of globalization, design of electronic hardware like intellectual property (IP) based system on chip (SoC) is not confined in a single site, but distributed across the globe [BHBN14, Boa05]. This involves development of indigenous IPs by third party

© The Author(s), under exclusive license to Springer Nature Switzerland AG 2021   69
K. Guha et al., *Self Aware Security for Real Time Task Schedules in Reconfigurable Hardware Platforms*, https://doi.org/10.1007/978-3-030-79701-0_4

IP (3PIP) vendors, who are physically distributed across the world. System developers procure IPs from such 3PIP vendors that meet the functional and performance criteria as specified by users, which are integrated to form the entire system. This design is sent to the foundry for fabrication which is again located in another location. After fabrication, related testing is performed, followed by packaging and distribution of the SoC. All such mechanisms is again handled by entities that are scattered worldwide. Possibility lies in the implantation of malicious codes in the IPs by untrusted 3PIP vendors [LRYK14, LRYK13]. Moreover, people in the foundry would have access to the unencrypted SoC design, which comprises all details like the IP blocks and interconnecting elements. Possibility remains in the fact that adversaries in the foundry may implant malicious circuitry in the IP during fabrication [XFT14]. Moreover, adversaries associated with testing and debugging the SoC have several attack options like reverse engineering the design and generating pirated copies of the design. Moreover, selling and distributing counterfeit components instead of original ones will result in degraded performance and malign the brand name of the producer [CD07, KPK08]. Though globalization has facilitated reduction of design cost and meet stringent marketing deadlines, but on the flip side, it has evicted the hardware root of trust.

Other than securing direct attacks on hardware, another aspect of hardware security involves ensuring reliable operations of the software stack, which encompasses protection of sensitive assets that are stored in hardware from malicious software and network attacks [BT18]. This deals with developing appropriate isolation between secure and insecure data and codes. Even separating and isolating users in a multi client system is essential. Thus, facilitating a trusted execution environment and protecting security critical assets for a system through proper isolation policies is the objective.

With the advent of industry 4.0 or fourth industrial revolution, focus is made on making the systems reconfigurable in nature [RB18]. This made system designers deploy reconfigurable hardware or field programmable gate arrays (FPGAs) in the system design and cloud service providers to deploy the same in the cloud platforms like Amazon's EC2 F1 services [Ser18], Microsoft's Project Catapult [CCP+16], etc. Such systems are associated with serving multiple users. Execution of user tasks is to be performed within strict deadlines and hence, real time task schedules are defined. However, hardware attacks may jeopardize such schedules [GSC18]. Attacks on FPGA based systems are two fold in nature. First, threats related to FPGA fabric and second, threats related to bitstreams or reconfigurable IPs may be associated. Erroneous results may be generated or denial of service may take place due to such vulnerability of hardware. Even excessive power dissipation may occur that may reduce lifetime of the system and cause an early system expiry. Previous works on hardware security do not focus on this issue.

This chapter is organized as follows. In Sect. 4.2, we discuss the difference between hardware security and hardware trust. Overview of hardware threats is dealt in Sect. 4.3. Section 4.4 discusses the lifecycle of an FPGA based system. Threats related to an FPGA based system, which can either be associated to FPGA bitstreams or FPGA fabric is presented in Sect. 4.5. Section 4.6 reviews the existing hardware security techniques in three classes, i.e. test time detection techniques, authentication mechanisms and runtime mitigation strategies. The present scope of this book is discussed in Sect. 4.7 and present chapter concludes in Sect. 4.8.

4.2 Overview of Hardware Threats

4.2.1 Hardware Trojan Horses (HTHs)

Malicious alterations made in system design by adversaries, that essentially remains in a dormant or inactive form during testing and initial phases of operation to avoid detection, but at runtime, gets triggered via some internal or external trigger to jeopardize operations. These can be termed as Hardware Trojan Horses [BHBN14, Boa05].

The name Hardware Trojans was derived from the great war of Troy, where it was difficult for the Greeks to enter the city of Troy due to its high walls surrounding the entire city, ensuring its security. The Greeks devised a plan where they built a huge wooden horse and gifted it to the Trojan army. However, the Greek soldiers stayed inside the wooden horse. The soldiers of Troy accepted the gift without knowing the malicious intent of the Greeks and took it into their secured city. This changed the course of the war as the Greek soldiers inside the seemingly trustworthy wooden horse came out at a certain pre-decided time and jeopardized the secure city of Troy. Similarly, in the present scenario also, the Hardware Trojan like the wooden horse in the city of Troy encapsulates malicious circuitry that cannot be detected previously (during testing), but triggers at a particular time to jeopardize the secure system like the secured city of Troy.

An HTH is particularly dangerous due to the fact that it evades detection in testing and initial stages of operation by remaining dormant [Boa05]. As illustrated in Fig. 4.1, a trigger module and a payload module are the essential components of an HTH architecture [BHBN14, TK10, RKK14]. The trigger may be either internal or external. A combinational or a sequential circuit with a pre-decided combination of node values as its activation criterion represents an internal trigger, while some sensors or antennas that have the ability to capture the triggering signals from some remote adversary represents an external trigger. Whent the trigger condition is satisfied, the payload which encapsulates the malicious functionality is triggered, which may jeopardize mission critical operations.

The US Government of Defense had even recognized HTHs as significant threats to mission critical applications in 2005 [Boa05].

Attacks may be active or passive in nature. Passive threats may not jeopardize a system and stop its operation, but causes leakage of secret information to an adversary via side channel parameters like delay, power, etc., i.e. affect the confidentiality of the system [LJM13, GSC15, GSC17a]. Active threats target the basic security primitives, integrity and availability of the system. Attack to integrity of the system may result in erroneous result generation or prevention of result generation suddenly

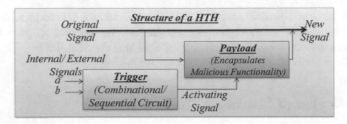

Fig. 4.1 Structure of HTH

at runtime [MWP+09, GSC19a, AAS14]. The basic security primitive, availability of a system is affected when a sudden, unintentional delay is caused, preventing result generation within deadline [GSC17b, GSC19b]. Real time systems are always associated with a deadline and finite or infinite delays caused by HTHs may jeopardize the system.

4.2.2 Piracy and Overbuilding

Unethical IP users or adversaries in an untrustworthy foundry may illegally duplicate and pirate an IP design without the permission of the IP designer. In this case, overbuilding of the IPs take place, with their distribution at low cost in the black market. This hampers the business of the genuine IP designers [RKM10, RKK14].

These affects business and causes enormous loss to genuine design houses. If in addition to IP piracy, HTHs are also implanted in them, then along with loss in business, these will also affect users by causing malfunctions during critical operations. These will lead to loss of reputation of the genuine design houses.

4.2.3 Reverse Engineering

In reverse engineering, an attacker tries to get acquainted with the intricate details of an IP. Reverse engineering can be either destructive or non-destructive in nature.

In destructive reverse engineering, an attacker tries to understand the intricate details of an IP by removing its package and finding the intricate connections within the IP. The original IP cannot be used after this, but with the knowledge gathered, the attacker can create several other IPs, which are functionally equivalent to the original IP and sell them at cheap prices in black market [TJ11].

While in non-destructive reverse engineering, intricate functional details is inspected based on side channel parameters like timing and power analysis of the intricate paths of the IP and reproducing them [Roh09]. Such attacks generate loss to the original IP design house.

Like IP piracy, this also mainly generate loss in business of the genuine design houses. Moreover, if HTHs are also implanted, then along with loss in business, malfunctions are caused during runtime affecting users and hampers the reputation of the genuine design houses.

4.2.4   Counterfeiting

Counterfeits are essentially components that are illegally reused or recycled and are associated with performance degradation at runtime [GDT14, GZFT14]. Degradation of performance related to aging is natural and is associated for all systems, after a certain time frame [HLK+15]. But use of counterfeit (or malicious) components in the design of an SoC may lead to induced aging. This will lead to delaying of task completion time over the normal time-frame and miss deadline. Like the previous case, no vulnerability can be detected during testing or initial phases of operation. However, at runtime, their performance degrades and prevents real time tasks to complete their operation within their respective deadlines. Thus, these threaten the availability of the system and pose active attacks. Issues related to counterfeiting is associated with an annual revenue loss of about $100 million [PT06]. Reports of Information Handling Services (IHS) shows that counterfeit parts have quadrupled since 2009 [Cas12]. In addition to hampering security, issues related to counterfeiting also affects reputation and reliability of a branded company.

Modern systems are mostly automatic and after deployment, they do not need supervision. Though it is feasible to monitor the runtime performance of a system and detect counterfeit components at runtime [KWJK14, ZXT12, SMAA13], but until and unless they are replaced, they continue to function [CMSW14]. Replacement of such components after deployment would be too expensive to perform.

Vulnerability may be introduced at any phase of system design. Trustworthiness of the third party vendors is an important issue. Common is the scenario where the 3PIP vendors may pose as an adversary and supply IPs with HTHs in them [LRYK13]. These HTHs are very hard to detect as the malicious functionality is represented via very few lines of HDL code. Even performing a 100% code coverage analysis is not sufficient to detect them [LJ01]. Moreover, the IPs may be supplied in an unreadable format to prevent duplicity and piracy. In such scenarios, it is very difficult to detect associated malwares. Moreover, as reference architectures of such customized IPs is unavailable, hence, detecting them by comparison with respect to side channel parameters is also infeasible. CAD tools for integrating the IPs is generally trusted, but the libraries and packages they use are also procured from third parties, and hence, trustworthiness is also a concern [VS16]. Adversaries in the foundry may insert malicious modules or HTHs in the empty spaces of the layout [XFT14]. Counterfeit components may also be inserted during manufacturing and packaging, i.e. in various stages of supply chain [CD07]. And as these remains in an inactive state during testing, hence, surpasses detection during the testing phases.

4.3   Hardware Trust and Hardware Security

4.3.1   Hardware Trust

Hardware Trust essentially involves issues that arise due to involvement of untrusted entities and third parties in the life cycle of hardware design. As a globalization technique is used for hardware design that involves several third party IP designers whose RTL designs are used and integrated in designing the entire system design, untrusted CAD tool vendors whose tools are used for RTL synthesis, several foundries for fabrication and several third party sites for testing, packaging and distribution [BT18].

Threat lies in the introduction of malicious codes supplied by third party IP vendors [LRYK13]. Untrusted CAD tools may comprise several malicious libraries or packages, the use of which may generate malicious designs after their synthesis, which if fabricated will generate malicious chips [BHBN14]. Adversaries in the foundries may implant malicious circuits during fabrication, especially in the unused portions of the layout [XFT14]. Agencies associated with testing may perform reverse engineering or side channel analysis to know the details of the circuitry and create pirated chips. The pirated chips do not perform to the required standard and causes increased power and energy dissipation and are even associated with runtime performance degradation [Guh20]. Even agencies associated with packaging and distribution can supply counterfeit products instead of the original ones, which will have a reduced lifetime and endure delays during runtime performance [GZFT14].

All such malicious elements introduced by adversaries during the hardware design causes issues related to trust or evicts the hardware root of trust.

4.3.2   Hardware Security

Hardware security defines detection strategies or mitigation techniques from unprecedented attacks, related to vulnerability of hardware [BHBN14, TK10]. Basically hardware security refers to countermeasures for attacks arising due to issues related to hardware trust like the malicious codes, malicious circuitry, malicious libraries and packages of the CAD tools that can cause active attacks like erroneous result generation and denial of service and passive attacks like leakage of secret information. Even counterfeit components causing unintentional delays at runtime can cause denial of service and is an active attack. Even pirated copies distributed may not function to the required standard and generate erroneous result or cause denial of service. Active attacks like erroneous result generation affects integrity of the system, while attacks related to denial of service impacts availability of the system [MWP+09, GSC19b]. Passive attacks like leakage of secret information hamper the confidentiality of the system [LJM13, GSC17a]. Moreover, these can even occur from lack of robust hardware support for softwares.

4.4 Life Cycle of FPGA Based System

Like a chip, an FPGA also has several phases in its development, which are not confined in a single site but outsourced to various geographic locations across the globe. Vulnerabilities associated in the various phases of its development are responsible for trust issues related to an FPGA based system. The diagrammatic representation of life cycle of FPGA based system is depicted in Fig. 4.2. The key entities in the development and working of an FPGA based system are:

4.4.1 Consumers

FPGAs find wide application in domains like automation, avionics, networking, defense products and consumer electronics. Consumers can either be companies that are involved in development of products in these domains or individuals. They may either procure standalone FPGAs directly from the FPGA vendors or provide custom specification to System Developers, which procure components like FPGAs, bitstreams, etc. and integrates them to form the final system and deliver the finished product to consumers.

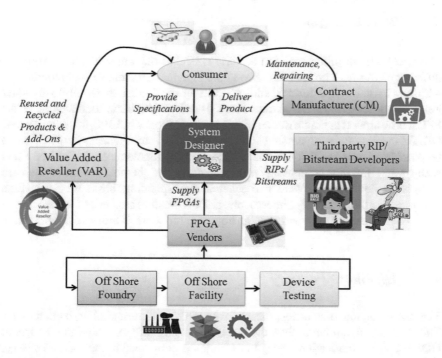

Fig. 4.2 Life cycle of FPGA based system

4.4.2   FPGA Based System Developer

On receiving specifications from consumers, they procure the various components like FPGAs, bitstreams, softwares, firmware, etc. from several third party vendors and integrate them to form the system as desired by the customer. The system developer may even deploy additional security features in the system to prevent IP theft via cloning or reverse engineering or tampering of the products. After required specifications are met, the system is delivered to the user.

4.4.3   Contract Manufacturer

For assembling, maintenance and repairing of the products, system developers may hire third parties, which are generally termed contract manufacturer. To facilitate contract manufacturers in their work of assembling, maintenance, testing and repairing of products, the system design is sent to them. They are even authorized to buy third party softwares, bitstreams and hardwares for their work, on behalf of the system designer.

4.4.4   FPGA Vendor

The FPGA vendor initiates design of various FPGAs that are suited to perform for different platforms. The lead FPGA vendors in the programmable logic market are Xilinx and Alterra. The former holds 45–50% of the market share, while the latter holds about 40–45% of the market share. It defines the basic architecture of the FPGA base array that comprises configurable logic blocks (CLBs), switch matrices, DRAM memory units, dynamic control module unit, etc. Present day FPGAs can be considered as a multi-functional chip, with programmable logic hardware and with several different computing components and peripherals. Layout of the base array is developed by the vendor to generate corresponding masks for fabrication. However, like the chip design houses, most FPGA vendors are fabless and the design is outsourced to other third party design houses or off-shore foundries for fabrication.

4.4.5   Off-Shore Foundry

The design and manufacturing process for an FPGA is almost similar to other chips. The off-shore foundries receives the layout of the base FPGA design in the form of GDS II files for fabrication. Design fabrication is performed in the foundries to get a raw fabricated base array.

4.4.6 Off-Shore Facility

The fabricated base array is outsourced again to another facility that is associated with packaging and assembly of an FPGA device. In general, to reduce manufacturing costs, such a facility is generally not a part of the foundry. This off-shore facility receives design details, device ID information, etc. from the FPGA vendor. These may be passed to another offshore facility for testing. Finally, these are sent to the FPGA vendor.

4.4.7 Third Party Reconfigurable IP/Bitstream Developers

An FPGA needs a reconfigurable IP or bitstream to configure its CLBs for a particular functional operation. Development of these bitstreams is associated with writing of hardware description language (HDL) codes, generally in VHDL or Verilog. Such HDL codes are synthesized in custom FPGA tools developed by the FPGA vendors, where several tests are performed like the logic test, design rule check, etc. After a code passes all the checks, it is synthesized to generate a bitstream that can be dumped on a specific family of FPGA boards to perform the function.

4.4.8 Value Added Reseller (VAR)

A VAR is not essential but in the present age where reuse and recycling has became quite prominent, VAR finds an important position. Products that are not aged and can be reused may be procured by system developers from VARs. Cost is less for such products than the original ones. Even pay per use licensing has became an essential feature in the present age, where several users may procure products for a certain number of uses and return them. For each use, a license is generated to activate the product. Payment is made as per the licenses procured. VARs are also useful for such systems. VARs have the responsibility to check the product before distributing it. Moreover, VARs are also associated with integrating confidential features that are unsharable with manufacturers like embedding cryptographic keys in memory of an FPGA. Since, providing the encrypted bitstream and key for decryption to a third party may lead to IP theft. To ensure safety, FPGA vendors must authorize selected third parties to act as a VAR, i.e. sell reused devices or facilitate pay per use licensing policy.

4.5 Overview of Threats Related to FPGA Based Systems

Threats can be related to bitstreams or FPGAs. Figure 4.3 provides a pictorial representation of overview of threats. These are also described as follows:

4.5.1 Attacks Related to Bitstreams

4.5.1.1 On Bitstreams

System designers do not have the expertise to build an entire system from scratch. Moreover, they are associated with stringent marketing deadlines. Hence, it is a common norm to procure different design components from third party vendors and integrate them to form the final design. Third party designers put in time and effort to generate bitstreams that serve to perform specific operations and solve different problems. Hence, bitstreams are generally considered as intellectual properties (IPs). These are essentially procured by system developers and integrated to form the system. However, if the logic of the bitstreams is revealed, then it can be duplicated and sold. This is an instance of IP theft. Moreover, if the contents of the bitstream are modified, then it will generate erroneous results, causing fatal consequences. Moreover, the name of the third party vendor will be maligned. Possible sources of attack related to the bitstream involve:

(i) System Designers/ CMs: Unethical system designers or contract manufacturers may create several types of attacks as discussed below:

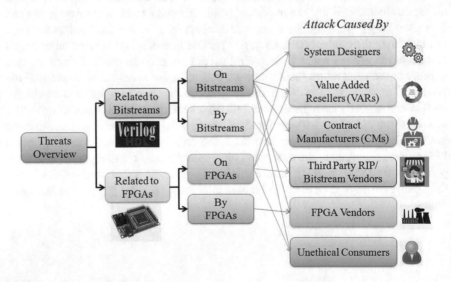

Fig. 4.3 Threats overview

Bitstream readback: JTAG port is mostly used as the programming interface for most FPGAs, i.e. for configuring the FPGA fabric. Several commands are present for testing and programming via the interface. Even commands exist that may cause retrieval of information bits from FPGA for verification of bitstream [JTE+12]. Hence, it is necessary to encrypt the bitstream to facilitate authenticate usage.

Bitstream Probing: As SRAM FPGAs are volatile in nature, hence, it is generally loaded during system power up, via the JTAG from an external memory like the flash memory. Hence, using an electrical probe, it is possible to intercept the bitstream transfer [DTB+15]. Even probing the memory of the FPGA may reveal the secret information used by the bitstreams.

Reverse Engineering: Deciphering the logic of the bitstream is the objective of reverse engineering [TJ11, Roh09]. This can be performed via brute force attack, where every possible input vector is provided to get a corresponding output. This can provide an idea of the bitstream logic. Moreover, via side channel analysis, i.e. analyzing the side channel parameters like power and delay for several operations can give the adversary an idea of the bitstream logic.

Bitstream Tampering: Individual bits can be altered by adversaries of a mapped bitstream by injecting faults [DWZZ13]. This may be created in a non-invasive or a semi-invasive format. In the former, no change in hardware takes place, while in the latter, a small change in hardware is associated. The former is associated with focused radiation and power adjustment. The latter is associated with mechanisms like optical fault injection with flashgun and laser pointer that changes individual bits of SRAM in a microcontroller. Moreover, direct modification of unencrypted bitstreams is also demonstrated [ZPT18].

(ii) VARs: If the system is reused or a pay per use licensing policy is adopted, then the system is passed through the VAR. The VAR then can launch similar attacks as that discussed for the system designer or CMs to get acquainted with the bitstream logic or tamper the bitstreams to malign the name of the third party bitstream developers.

(iii) FPGA Vendors (Adversaries in the Foundry): Adversaries in the foundry may implant malicious circuitry of HTHs that may snoop the bitstream workings and reveal secret information via covert side channels to adversaries. Other than such passive attacks, implanted HTH in the FPGAs may also disrupt smooth operations of the bitstreams by causing unintentional delays or even by stopping their operations.

(iv) Unethical Users: Attacks like reverse engineering, bitstream probing can also be performed by unethical users to understand the bitstream logic.

4.5.1.2 By Bitstreams

Third party vendors who design the bitstreams may also act as adversaries and implant malicious codes during generation of the bitstreams. Such malicious codes are difficult to detect as these are of a few lines embedded in a sea of original ones. These are also termed as HTHs. These basically snoop operations of the user and reveal secret information via covert side channels. Even if customized FPGAs are procured by the vendors for specific operations, such bitstreams have the capability to understand the underlying FPGA hardware logic and reveal them via covert side channels. Moreover, these bitstreams may take control of the entire system if proper authentication facility is absent.

4.5.2 Attacks Related to FPGAs

4.5.2.1 On FPGAs

Different types of functionalities are performed by reconfiguring the FPGA fabric or base hardware with related bitstreams, at different instance of time. However, if the FPGA fabric gets compromised, adversaries will be able to get control of the system. Some of the ways in which the FPGA fabric may get compromised are:

(i) Malicious 3PIP vendors: Malicious third party vendors who design the bitstreams may design them in such a manner that these may leak secret information stored in memory like the secret keys of cryptographic operations or may even take control of the system. Hence, without proper authentication, giving access to bitstreams for re-configuring an FPGA is undesirable.

(ii) System Designers/ VARs/ CMs: For specific applications, some FPGAs may be designed in a customized manner by the FPGA vendors. However, unethical system designers or VARs or CMs may perform various operations like reverse engineering and probing attacks on the FPGAs to get acquainted with the custom hardware. Based on the received information, they may create duplicate customized FPGA hardware and market them at lower price. Such an attack can be considered as an IP theft, leading to piracy.

(iii) Unethical users/consumers: Unethical users or consumers may also use reverse engineering and probing attacks to get acquainted with custom FPGA logic and secret information stored in memory. Then they may generate pirated copies and sell them.

4.5.2.2 By FPGA Fabric

System designers may order FPGA vendors to produce customized FPGAs for specific applications or may also procure generalized FPGAs from FPGA vendors. As discussed previously, design of an FPGA is not enacted in a single site, but outsourced to several offshore facilities. Adversaries in such facilities may create several attack scenarios as discussed:

(i) Foundries: During fabrication, malicious circuitry or HTHs may be implanted in the FPGA fabric by adversaries in the foundry. These may be implanted in the unused portions of the fabric, so that these do not get detected easily. In general, HTHs stay dormant during testing and initial phases of operation, but get activated after a certain trigger condition is satisfied. The trigger may be either internal, i.e. a rare combination of values preset by the adversary or external, i.e. via signals sent by the adversary and received over inbuilt antenna and sensors. Such HTHs may stop the operations of the FPGA at runtime, or may cause generation of erroneous results or may cause unintentional delays in operation that prevents real time tasks to complete within deadline or may also leak secret information of users via covert side channels. Thus, all the basic security primitives, i.e. confidentiality, integrity and availability may get affected.

(ii) Off-shore Facilities: Off-shore facilities that are associated with important mechanisms like testing may skip essential tests, which may lead to malfunction at runtime. Erroneous results may be generated and systems may stop function during a deadlock scenario. These are active attacks, i.e. affect the integrity and availability of the system and may also cause fatal consequences.

4.6 Overview of Hardware Security Techniques for FPGA Based Systems

Securing from the vulnerability of hardware is difficult due to the following factors:

(i) Inherent opaqueness of the supplied IPs by third party vendors is a hurdle for detection of modified components. The destructive tests and reverse engineering strategies are quite slow, as well as expensive.

(ii) Technology scaling to the intrinsic limits of physical semiconductor device and mask imprecisions creates non-determinism and inherent difficulty to differentiate between process variations of side channel parameters and operations of malicious entities like HTHs.

(iii) As location of HTH is unknown in the huge system space, hence, difficulty arises for testing of appropriate nodes.

(iv) As traditional post manufacturing testing is ineffective, different strategies were developed to counteract HTHs [BHBN14].

We have categorized these into three parts, (i) Test Time Detection Techniques, (ii) Protection Based on Authentication and (iii) Runtime Mitigation Mechanisms. Figure 4.4 depicts the same. Detailed description of the existing techniques are as follows:

4.6.1 Test Time Detection Techniques

4.6.1.1 Logic Testing

In logic testing, typical test vectors are generated in such a manner so that they can instigate and trigger the malicious effect of an HTH. Authors in [CWP+09, LJYM12] propose a mechanism to detect low probability conditions at the various internal nodes of an IP design and generate optimal vectors that has the capability to activate the low probability malicious nodes individually, based on their logic values.

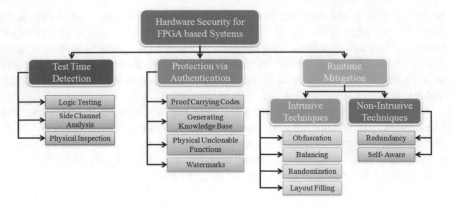

Fig. 4.4 Overview of hardware security techniques for FPGA based systems

Though the mechanism is apt for small architectures with a limited number of peripherals, problem arises when SoC complexity increases, i.e. the number of internal nodes, related functionality and number of peripherals increases.

4.6.1.2   Side Channel Analysis

Side channel analysis ensures security of the experimental model with respect to a standard or golden model, based on side channel parameters like power, leakage current, delay, etc. A plethora of strategies exist, some of which are discussed here. Path delay based side channel analysis is proposed in [RL09, YM08]. Side channel analysis based on static current is exhibited in [AARP10]. Side channel analysis based on transient current is also performed in [Wil02] and [BH09]. The former is associated with region-based partitioning, while the latter uses directed test vector generation. Use of multiple parameters for side channel analysis is presented in [NDC+13].

However, when the size of the experimental model is large and the HTH size is comparatively negligible, such side channel based malware detection methodologies cease to be effective. Even obtaining some reference or golden models are difficult for customized IP designs.

For practical scenarios, to maximize the level of confidence for test time detection strategies, a combination of logic testing and side channel analysis is performed.

4.6.1.3   Physical Inspection

Physical inspection may be carried out via techniques like X-ray analysis [MJN+10]. These are effective for analysis of vulnerabilities associated with FPGA devices. However, these are quite time consuming and costly.

4.6.2 Protection via Authentication

4.6.2.1 Proof Carrying Codes (PCC)

Formulation of a set of security related properties is carried out, and formal proof of these is implanted by the designer in the supplied IP. Any undesired modification to the IP also changes these proofs and is detected by the user [LJYM12].

4.6.2.2 Generating Knowledge Base (KB)

Analysis via expert-system-based mechanism is proposed in [SSW+14b, SSW+14a]. In these, knowledge about designs and their related specifications are stored in a KB. KB rules are formulated, which are used to perform static and dynamic analysis. However, building a KB is quite challenging which involves building of specification ontologies, property-based models, rule-base and related assertion libraries.

4.6.2.3 Physical Unclonable Functions (PUFs)

Side channel parameters of a physical device like timing or delay depend on the intrinsic property of the base semiconductor material of the device. These never match with another device. This property was first utilized by Pappu et.al to generate PUFs [PRTG02]. Generation of PUFs for authentication and generation of secret keys was shown in [SD07]. Generation of digital PUFs for authentication of reconfigurable hardware devices was exhibited in [XP14]. A mechanism for self measuring the combinatorial circuit delays was depicted in [WSC09] and based on this mechanism, a PUF based time bounded authentication strategy for FPGAs was developed in [MK11].

4.6.2.4 Watermarking

Watermarking can be considered a strategy where an IP vendor or the source designer embeds unique signature for the user to verify and authenticate its source of origin. Watermarking in hard IPs for pre-fabrication and post-fabrication verification was proposed in [SS15]. Embedding of watermarks in reconfigurable scan architecture for ensuring authentication of IPs in SoC was also exhibited in [SS16]. Watermarking can also be facilitated indirectly like using right most bits of timing constraints on signal delays of constraint file [JYPQ03] or via side channel parameters like power traces, as shown in [ZT08].

Authentication techniques are generally applied to detect fake components. However, as discussed, these can also be used for detection of components with HTHs.

4.6.3   Runtime Mitigation Mechanisms

Ensuring security at runtime is often termed as the last line of defense by eminent minds of the present arena [BHBN14]. Such runtime security techniques can be classified into two types: (i) intrusive techniques (ii) non-intrusive techniques.

4.6.3.1   Intrusive Mechanisms

Intrusive security mechanisms is associated with implantation of security codes in the IP designs. This is performed by the designers during the hardware description language (HDL) phase of IP design. These involve:

Obfuscation Techniques: In this mechanism, additional circuits are incorporated to obfuscate the functional and structural properties of an IP design [CB09, CB11]. This makes a circuit operate or perform in two different modes, (i) normal mode and (ii) obfuscated mode. Normal mode operations is carried out only if accurate keys are provided by the user, else operation is performed in the obfuscated mode, where actual result is not generated.

Balancing/Randomization Techniques: The designer either balances the IP design in such a manner that difference in side channel parameters of the various paths of the IPs is mitigated [AZM10, JIP17]. When an attacker implants malicious elements in any of the paths of the IP design, detection is possible via side channel analysis. The designer may also implant additional elements in the design to randomize the side channel parameters, which will confuse the adversaries [THAC18].

Filler Techniques: This mechanism involves filling of additional spaces of a layout by implanting a built in self authentication circuit [XFT14]. Not only authentication tests are performed by such a circuitry, but also implantation of HTHs in the untrustworthy foundries are avoided as the un-utilized spaces are used up.

4.6.3.2   Non-intrusive Mechanisms

In non-intrusive mechanisms, security is provided without modification or tampering of IPs that are procured from various third party IP vendors. This is essentially important in scenarios where the SoC integrator is unaware of the internal design of the procured IPs. In addition to this, the supplied IPs may be securely packaged or be in an unreadable format to prevent code stealing and prevent unauthorized replication of IP design. The SoC integrator is only able to access the supplied IPs via the IP peripherals.

Redundancy Based Strategies: In redundancy based security mechanism, multiple IPs that perform the same functional operation is procured from a variety of sources. A task is simultaneously executed in all of them. After completion of task execution, a comparison [MWP+09] or polling strategy [AAS14] is used to determine and generate the correct result.

However, increase in the number of vendors increases cost. Vendor optimization strategies are proposed in [SBM17, RSK16]. However, such a methodology has high area and power overhead, with low throughput.

Self Aware Strategies: In these, adherence is made to Observe-Decide-Act policy [GSC17b, GSC19b, GSC17a]. The runtime performance is monitored continuously. On detecting a deviation from normal scenario, course of action is deciphered. Based on that, necessary action is taken to mitigate the vulnerability.

4.7 Present Scope

For the present scope of this book, we consider and analyze threats that may occur to real time task schedules, due to the vulnerability of hardware.

Existing literature is silent on hardware threats to real time task schedules. Hence, related mitigation strategies for real time task schedules from the vulnerability of hardware are also missing in current literature.

We focus only on runtime security strategies. We analyze the limitations of redundancy based strategies and propose how security can be ensured via the self aware approach.

4.8 Conclusion

In this chapter, we provide a brief overview of hardware threats, followed by hardware trust and hardware security. Then life cycle of a FPGA based system is discussed, along with related threats. Finally, hardware security techniques for a FPGA based system is discussed. However, such FPGA based systems are associated with serving multiple users or clients at the same time. Hence, the FPGAs are partitioned into multiple virtual portions (VPs), where independent task execution can take place. Thus, virtualization is followed, where the FPGAs are spatially and temporarily shared among many users. For this, real time task schedules need to be be developed offline, which is followed at runtime. But threats related to hardware may jeopardize such real time task schedules. Existing works does not focus on such threats, nor provides security strategies to mitigate such issues. Analyzing passive and active threats for such scenarios and proposing self aware security strategies is the present objective.

References

[AARP10]   J. Aarestad, D. Acharyya, R. Rad, J. Plusquellic, Detecting trojans through leakage current analysis using multiple supply padI_{DDQ}s. IEEE Trans. Inf. Forensics Secur. **5**(4), 893–904 (2010)

[AAS14]   H.A.M. Amin, Y. Alkabani, G.M.I. Selim, System-level protection and hardware trojan detection using weighted voting. J. Adv. Res. **5**(4), 499–505 (2014)

[AZM10]   A. Askarov, D. Zhang, A.C. Myers, Predictive black-box mitigation of timing channels, in *Proceedings of the 17th ACM Conference on Computer and Communications Security* (CCS '10, 2010) pp. 297–307

[BH09]   M. Banga, M.S. Hsiao, A novel sustained vector technique for the detection of hardware trojans, in *2009 22nd International Conference on VLSI Design* (2009), pp. 327–332

[BHBN14]   S. Bhunia, M.S. Hsiao, M. Banga, S. Narasimhan, Hardware trojan attacks: threat analysis and countermeasures. Proc. IEEE **102**(8), 1229–1247 (2014)

[Boa05]   D.S. Board, Task force on high performance microchip supply (2005). http://www. acq.osd.mil/dsb/reports/ADA435563.pdf

[BT18]   S. Bhunia, M. Tehranipoor, *Hardware Security—A Hands on Approach* (Elsevier Morgan Kaufmann Publishers, 2018). ISBN: 9780128124772

[Cas12]   J. Cassell, Reports of counterfeit parts quadruple since 2009, challenging US defense industry and national security (2012). http://news.ihsmarkit.com/press-release/design-supply-chain/reports-counterfeit-parts-quadruple-2009-challenging-us-defense-in

[CB09]   R.S. Chakraborty, S. Bhunia, HARPOON: an obfuscation-based SoC design methodology for hardware protection. IEEE Trans. Comput. Aided Des. Integr. Circuits Syst. **28**(10), 1493–1502 (2009)

[CB11]   R.S. Chakraborty, S. Bhunia, Security against hardware trojan attacks using key-based design obfuscation. J. Electron. Test. **27**(6), 767–785 (2011)

[CCP+16]   A.M. Caulfield, E.S. Chung, A. Putnam, H. Angepat, J. Fowers, M. Haselman, S. Heil, M. Humphrey, P. Kaur, J.-Y. Kim, D. Lo, T. Massengill, K. Ovtcharov, M. Papamichael, L. Woods, S. Lanka, D. Chiou, D. Burger, A cloud-scale acceleration architecture, in *2016 49th Annual IEEE/ACM International Symposium on Microarchitecture (MICRO)* (2016), pp. 1–13

[CD07]   K. Chatterjee, D. Das, Semiconductor manufacturers' efforts to improve trust in the electronic part supply chain. IEEE Trans. Compon. Packag. Technol. **30**(3), 547–549 (2007)

[CMSW14]   X. Cui, K. Ma, L. Shi, K. Wu, High-level synthesis for run-time hardware trojan detection and recovery, in *Proceedings of the 51st Annual Design Automation Conference* (DAC '14, 2014), pp. 157:1–157:6

[CWP+09]   R.S. Chakraborty, F. Wolff, S. Paul, C. Papachristou, S. Bhunia, MERO: a statistical approach for hardware trojan detection, in *Proceedings of the 11th International Workshop on Cryptographic Hardware and Embedded Systems* (CHES '09, 2009), pp. 396–410

[DTB+15]   R. Druyer, L. Torres, P. Benoit, P.V. Bonzom, P. Le-Quere, A survey on security features in modern FPGAs, in *2015 10th International Symposium on Reconfigurable Communication-centric Systems-on-Chip (ReCoSoC)* (2015), pp. 1–8

[DWZZ13]   Z. Ding, W. Qiang, Y. Zhang, L. Zhu, Deriving an NCD file from an FPGA bitstream: methodology, architecture and evaluation. Microprocess. Microsyst. **37**(3), 299–312 (2013)

[GDT14]   U. Guin, D. DiMase, M. Tehranipoor, Counterfeit integrated circuits: detection, avoidance, and the challenges ahead. J. Electron. Test. **30**(1), 9–23 (2014)

[GMSC20]   K. Guha, A. Majumder, D. Saha, A. Chakrabarti, Ensuring green computing in reconfigurable hardware based cloud platforms from hardware trojan attacks, in *2020 IEEE REGION 10 CONFERENCE (TENCON)* (2020), pp. 1380–1385

[GSC15]   K. Guha, D. Saha, A. Chakrabarti, RTNA: securing SOC architectures from confi-
           dentiality attacks at runtime using ART1 neural networks, in *2015 19th International
           Symposium on VLSI Design and Test* (2015), pp. 1–6
[GSC17a]   K. Guha, D. Saha, A. Chakrabarti, Real-time SoC security against passive threats
           using crypsis behavior of geckos. J. Emerg. Technol. Comput. Syst. **13**(3), 41:1–41:26
           (2017)
[GSC17b]   K. Guha, D. Saha, A. Chakrabarti, Self aware SoC security to counteract delay inducing
           hardware trojans at runtime, in *2017 30th International Conference on VLSI Design
           and 2017 16th International Conference on Embedded Systems (VLSID)* (2017), pp.
           417–422
[GSC18]   K. Guha, S. Saha, A. Chakrabarti, Shirt (self healing intelligent real time) schedul-
           ing for secure embedded task processing, in *2018 31st International Conference on
           VLSI Design and 2018 17th International Conference on Embedded Systems (VLSID)*
           (2018), pp. 463–464
[GSC19a]   K. Guha, D. Saha, A. Chakrabarti, SARP: self aware runtime protection against
           integrity attacks of hardware trojans, in *VLSI Design and Test* (Singapore, 2019),
           pp. 198–209
[GSC19b]   K. Guha, D. Saha, A. Chakrabarti, Stigmergy-based security for SoC operations from
           runtime performance degradation of SoC components. ACM Trans. Embed. Comput.
           Syst. **18**(2), 14:1–14:26 (2019)
[Guh20]   A. Majumder, D. Saha, A. Chakrabarti, K. Guha, Dynamic power-aware scheduling
           of real-time tasks for FPGA-based cyber physical systems against power draining
           hardware trojan attacks. J. Supercomput. **76**(11), 8972–9009 (2020)
[GZFT14]   U. Guin, X. Zhang, D. Forte, M. Tehranipoor, Low-cost On-chip structures for com-
           bating die and IC recycling, in *Proceedings of the 51st Annual Design Automation
           Conference* (DAC '14, 2014), pp. 87:1–87:6
[HLK+15]   K. Huang, Y. Liu, N. Korolija, J.M. Carulli, Y. Makris, Recycled IC detection based
           on statistical methods. IEEE Trans. Comput.-Aided Des. Integr. Circuits Syst. **31**(6),
           947–960 (2015)
[JIP17]   D. Jayasinghe, A. Ignjatovic, S. Parameswaran, NORA: algorithmic balancing without
           pre-charge to thwart power analysis attacks, in *2017 30th International Conference on
           VLSI Design and 2017 16th International Conference on Embedded Systems (VLSID)*
           (2017), pp. 167–172
[JTE+12]   K. Jozwik, H. Tomiyama, M. Edahiro, S. Honda, H. Takada, Comparison of preemp-
           tion schemes for partially reconfigurable FPGAs. IEEE Embed. Syst. Lett. **4**(2), 45–48
           (2012)
[JYPQ03]   A.K. Jain, L. Yuan, P.R. Pari, G. Qu, Zero overhead watermarking technique for FPGA
           designs, in *Proceedings of the 13th ACM Great Lakes Symposium on VLSI* (GLSVLSI
           '03, 2003), pp. 147–152
[KPK08]   T.H. Kim, R. Persaud, C.H. Kim, Silicon odometer: an on-chip reliability monitor
           for measuring frequency degradation of digital circuits. IEEE J. Solid-State Circuits
           43(4), 874–880 (2008)
[KWJK14]   J. Keane, X. Wang, P. Jain, C.H. Kim, On-chip silicon odometers for circuit aging
           characterization, in *Bias Temperature Instability for Devices and Circuits* (2014), pp.
           679–717
[LJ01]   C. Liu, J. Jou, Efficient coverage analysis metric for HDL design validation. IEE
           Proc.-Comput. Digit. Tech. **148**(1), 1–6 (2001)
[LJM13]   Y. Liu, Y. Jin, Y. Makris, Hardware trojans in wireless cryptographic ICs: silicon
           demonstration & detection method evaluation, in *Proceedings of the International
           Conference on Computer-Aided Design* (ICCAD '13, 2013), pp. 399–404
[LJYM12]   E. Love, Y. Jin, Y.Y. Makris, Proof-carrying hardware intellectual property: a pathway
           to trusted module acquisition. IEEE Trans. Inf. Forensics Secur. **7**(1), 25–40 (2012)

[LRYK13]   C. Liu, J. Rajendran, C. Yang, R. Karri, Shielding heterogeneous MPSoCs from untrustworthy 3PIPs through security-driven task scheduling, in *2013 IEEE International Symposium on Defect and Fault Tolerance in VLSI and Nanotechnology Systems (DFTS)* (2013), pp. 101–106

[LRYK14]   C. Liu, J. Rajendran, C. Yang, R. Karri, Shielding heterogeneous MPSoCs from untrustworthy 3PIPs through security- driven task scheduling. IEEE Trans. Emerg. Top. Comput. **2**(4), 461–472 (2014)

[MJN+10]   T. Madden, P. Jemian, S. Narayanan, A. Sandy, M. Sikorski, M. Sprung, J. Weizeorick, Fpga-based compression of streaming x-ray photon correlation spectroscopy data, in *IEEE Nuclear Science Symposuim Medical Imaging Conference* (2010), pp. 730–733

[MK11]   M. Majzoobi, F. Koushanfar, Time-bounded authentication of FPGAs. IEEE Trans. Inf. Forensics Secur. **6**(3), 1123–1135 (2011)

[MWP+09]   D. McIntyre, F. Wolff, C. Papachristou, S. Bhunia, D. Weyer, Dynamic evaluation of hardware trust, in *2009 IEEE International Workshop on Hardware-Oriented Security and Trust* (2009), pp. 108–111

[NDC+13]   S. Narasimhan, D. Du, R.S. Chakraborty, S. Paul, F.G. Wolff, C.A. Papachristou, K. Roy, S. Bhunia, Hardware trojan detection by multiple-parameter side-channel analysis. IEEE Trans. Comput. **62**(11), 2183–2195 (2013)

[PRTG02]   R. Pappu, B. Recht, J. Taylor, N. Gershenfeld, Physical one-way functions. Science **297**(5589), 2026–2030 (2002)

[PT06]   M. Pecht, S. Tiku, Bogus: electronic manufacturing and consumers confront a rising tide of counterfeit electronics. IEEE Spectr. **43**(5), 37–46 (2006)

[RB18]   L. Ribeiro, M. Björkman, Transitioning from standard automation solutions to cyber-physical production systems: an assessment of critical conceptual and technical challenges. IEEE Syst. J. **12**(4), 3816–3827 (2018)

[RKK14]   M. Rostami, F. Koushanfar, R. Karri, A primer on hardware security: models, methods, and metrics. Proc. IEEE **102**(8), 1283–1295 (2014)

[RKM10]   J.A. Roy, F. Koushanfar, I.L. Markov, Ending piracy of integrated circuits. Computer **43**(10), 30–38 (2010)

[RL09]   D. Rai, J. Lach, Performance of delay-based trojan detection techniques under parameter variations, in *2009 IEEE International Workshop on Hardware-Oriented Security and Trust* (2009), pp. 58–65

[Roh09]   P. Rohatgi, *Improved Techniques for Side-Channel Analysis* (Springer US, Boston, MA, 2009), pp. 381–406

[RSK16]   J.J. Rajendran, O. Sinanoglu, R. Karri, Building trustworthy systems using untrusted components: a high-level synthesis approach. IEEE Trans. Very Large Scale Integr. (VLSI) Syst. **24**(9), 2946–2959 (2016)

[SBM17]   A. Sengupta, S. Bhadauria, S.P. Mohanty, TL-HLS: methodology for low cost hardware trojan security aware scheduling with optimal loop unrolling factor during high level synthesis. IEEE Trans. Comput.-Aided Des. Integr. Circuits Syst. **36**(4), 655–668 (2017)

[SD07]   G.E. Suh, S. Devadas, Physical unclonable functions for device authentication and secret key generation, in *Proceedings of the 44th Annual Design Automation Conference* (DAC '07, 2007), pp. 9–14

[Ser18]   Amazon Web Services. Amazon Elastic Compute Cloud User Guide (2018). https://docs.aws.amazon.com/AWSEC2/latest/UserGuide/ec2-ug.pdf

[SMAA13]   H.G. Stratigopoulos, S. Mir, L. Abdallah, J. Altet, Defect-oriented Non-intrusive RF test using on-chip temperature sensors, in *Proceedings of the 2013 IEEE 31st VLSI Test Symposium* (VTS '13, 2013), pp. 1–6

[SS15]   D. Saha, S. Sur-Kolay, Watermarking in hard intellectual property for pre-fab and post-fab verification. IEEE Trans. Very Large Scale Integr. (VLSI) Syst. **23**(5), 801–809 (2015)

[SS16]   D. Saha, S. Sur-Kolay, Embedding of signatures in reconfigurable scan architecture for authentication of intellectual properties in system-on-chip. IET Comput. Digit. Tech. **10**(3), 110–118 (2016)

[SSW+14a] B. Singh, A. Shankar, F. Wolff, C. Papachristou, D. Weyer, S. Clay, Cross-correlation of specification and RTL for soft IP analysis, in *2014 Design, Automation Test in Europe Conference Exhibition (DATE)* (2014), pp. 1–6

[SSW+14b] B. Singh, A. Shankar, F. Wolff, D. Weyer, C. Papachristou, B. Negi, Knowledge-guided methodology for third-party soft IP analysis, in *2014 27th International Conference on VLSI Design and 2014 13th International Conference on Embedded Systems* (2014), pp. 246–251

[THAC18] D. Trilla, C. Hernandez, J. Abella, F.J. Cazorla, Cache side-channel attacks and time-predictability in high-performance critical real-time systems, in *Proceedings of the 55th Annual Design Automation Conference* (DAC '18, ACM, New York, NY, USA 2018), pp. 98:1–98:6

[TJ11] R. Torrance, D. James, The state-of-the-art in semiconductor reverse engineering, in *2011 48th ACM/EDAC/IEEE Design Automation Conference (DAC)* (2011), pp. 333–338

[TK10] M. Tehranipoor, F. Koushanfar, A survey of hardware trojan taxonomy and detection. IEEE Des. Test Comput. **27**(1), 10–25 (2010)

[VS16] N. Veeranna, B. Schafer, Hardware trojan detection in behavioral intellectual properties(IPs) using property checking techniques. IEEE Trans. Emerg. Top. Comput. (99), 1 (2016)

[Wil02] M.A. Williams, Anti-trojan, trojan detection with in-kernel digital signature testing of executables, in *Security Software Engineering: NetXSecure NZ Limited, Technical Reports* (2002), pp. 1–12

[WSC09] J.S.J. Wong, P. Sedcole, P.Y.K. Cheung, Self-measurement of combinatorial circuit delays in FPGAs. ACM Trans. Reconfigurable Technol. Syst. **2**(2), 10:1–10:22 (2009)

[XFT14] K. Xiao, D. Forte, M. Tehranipoor, A novel built-in self-authentication technique to prevent inserting hardware trojans. IEEE Trans. Comput.-Aided Des. Integr. Circuits Syst. **33**(12), 1778–1791 (2014)

[XP14] T. Xu, M. Potkonjak, Robust and flexible FPGA-based digital PUF, in *2014 24th International Conference on Field Programmable Logic and Applications (FPL)* (2014), pp. 1–6

[YM08] Y. Jin, Y. Makris, Hardware trojan detection using path delay fingerprint, in *2008 IEEE International Workshop on Hardware-Oriented Security and Trust* (2008), pp. 51–57

[ZPT18] D. Ziener, J. Pirkl, J. Teich, Configuration tampering of BRAM-based AES implementations on FOGAs, in *2018 International Conference on ReConFigurable Computing and FPGAs (ReConFig)* (2018), pp. 1–7

[ZT08] D. Ziener, J. Teich, Power signature watermarking of IP cores for FPGAs. J. Signal Process. Syst. **51**(1), 123–136 (2008)

[ZXT12] X. Zhang, K. Xiao, M. Tehranipoor, Path-delay fingerprinting for identification of recovered ICs, in *2012 IEEE Int. Symp. on Defect and Fault Tolerance in VLSI and Nanotechnology Systems* (2012), pp. 13–18

Chapter 5
Bypassing Passive Attacks

5.1 Introduction

Passive is the threat when system confidentiality is at stake. Such threats do not cause direct damage by jeopardizing operations and generating erroneous results or causing stoppage of operations or even does not delay real time operations to cause a deadline miss. Hence, such threats are passive in nature. This involves leakage of secret information to adversaries [LJM13, GSC17a, GSC15]. For example, the secret key that is associated with cryptographic operations by a genuine user may be leaked to an adversary

Such attacks are generally performed via various side channel parameters like power, delay, leakage current, etc. [NDC+13, AARP10]. The side channel parameters are monitored by adversaries over a range of operations to get acquainted with the secret information. A single security module that can monitor all possible side channel parameters and counteract leakage of secret information via all such channels is practically infeasible, as different types of monitors are required to monitor different side channel parameters. Moreover, if such a security module is even constructed, then the overhead induced will be tremendous with respect to area, power and cost. Hence, it is wise to develop security modules that is associated with monitoring and mitigation based on a particular type of side channel parameter, as important for the application scenario.

The present scenario involves deployment of FPGAs in environments like IoTs and cloud environments [RB18], where tasks from multiple users are executed in a suitable time frame. To ensure suitable sharing of the FPGA resource, both with respect to space and time, the FPGA is divided into various virtual portions that execute tasks independently with the aid of bitstreams or reconfigurable intellectual properties (RIP) procured from various third party IP vendors [GSC18, Guh20]. Real time task schedules are even generated offline, based on which task executions are performed online that ensures fair sharing of execution time among different users and prevents interference during operation.

© The Author(s), under exclusive license to Springer Nature Switzerland AG 2021
K. Guha et al., *Self Aware Security for Real Time Task Schedules in Reconfigurable Hardware Platforms*, https://doi.org/10.1007/978-3-030-79701-0_5

However, different bitstreams or RIPs may take different amount of times for execution of a particular task, as the algorithm used by different vendors will be different in carrying out execution of the same task. Hence, the real time schedules dedicate space and time for worst case execution time for executing a task. Moreover, it is customary for the designer to allot some amount of buffer time, in case some finite delay occurs at runtime. This additional time may be utilized to cause differential delay in generating the results and facilitating knowledge of secret information by adversaries [GSC15, GSC17a].

Thus, for the present case, secret information is generally revealed via the side channel parameter, timing, by causing differential delay in result generation as per the secret key over a range of operations. Use of other side channel parameters like power analysis, leakage current analysis, etc. is not quite effective in the present scenario as several tasks are executed at the same time and monitoring the entire system power or the leakage current for the FPGA based system will not aid the adversary, who is interested in side channel parameter analysis for a particular task operation.

Existing security mechanisms against confidentiality attacks essentially focus on generating confusion or randomness in operations like obfuscating or algorithm balancing or masking to bypass the threats [AZM10, JIP17]. Even such strategies adhere to golden models for detection of vulnerability. However, they do not focus on securing task operations of real time task schedules. Moreover, most of the strategies are associated in the pre-deployment phase or ensures security via an intrusive mechanism as the security elements are implanted in the design. These do not consider bypassing such threats at runtime via a self-aware mechanism or via a non-intrusive strategy. In addition to this, securing confidentiality attacks from vulnerability associated with procured RIPs or FPGA devices from third party vendors is also not focused.

In the present chapter, we discuss how a particular task in a real time task schedule may be associated with confidentiality attack via the side channel parameter, timing. The threat may arise either due to vulnerability associated with the procured RIPs or the FPGA device. We describe the development of self-aware agents and demonstrate how these may bypass such threats at runtime via a non-intrusive mechanism. Finally, experimentation is performed with real time cryptographic operations and related results are presented.

This chapter is organized as follows. Section 5.2 deals with the system model, while the threat model is described in Sect. 5.3. Section 5.4 presents the self-aware security mechanism. Experimentation and results are depicted in Sect. 5.5 and the chapter concludes in Sect. 5.6.

5.2 System Model

We consider a multi user system with either a single FPGA resource or a number of FPGA resources. Users send their tasks, which are collected by the scheduler and allocated to the FPGA VPs as per the real time schedules designed offline. Signals are given by the scheduler to the partial reconfiguration (PR) module at appropriate instants of time that re-configures the FPGA VPs via the Internal Configuration Access Port (ICAP), with bitstreams or RIPs procured from third party vendors stored in the memory, as depicted in Fig. 5.1.

The FPGAs may either perform in a fully re-configurable mode or in a partially re-configurable mode. Details of scheduling for either mode are discussed in previous chapters. However, for quick recapitulation, we will discuss them briefly over here.

A set of tasks is considered that needs to be executed in an FPGA, with a certain number of VPs. The scheduling strategy initially determines the work share for all tasks in a time slice. The length of a time slice is equivalent to the minimum deadline among all the tasks. Residual work shares are to be completed in the following time slices. A task may comprise of one or more subtasks. Each subtask execution is associated with re-configuring a particular VP with a procured bitstream from a 3PIP vendor for a certain time instant, during which its execution is carried out. Then the VP can be reconfigured with another bitstream for execution of another subtask. Placement of the subtasks is performed in an appropriate manner in all the available VPs for a time slice in such a manner that none of the tasks miss their deadline.

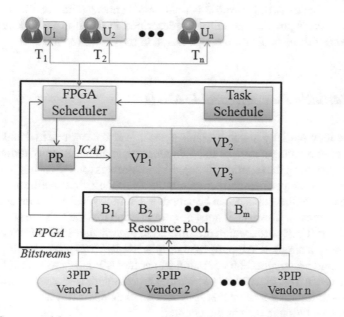

Fig. 5.1 System model

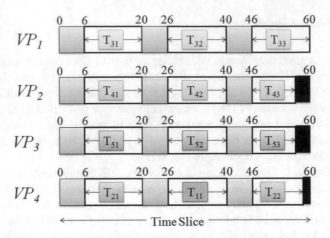

Fig. 5.2 Task schedule for a full re-configurable system

5.2.1 Fully Re-configurable Mode

Scheduling for a fully re-configurable system is described in detail in Part 2 Chap. 3. In this, reconfiguration of all the VPs during a time slice are carried out simultaneously with bitstreams for performing a particular subtask. If the subtask completes early, then the remaining time can be considered as slack and for the present scenario, the slack remains unused. Figure 5.2 is a pictorial representation of a task schedule for a fully re-configurable system that is described in detail in Part 2 Chap. 3. We take the aid of this example for demonstration in the current context.

5.2.2 Partially Re-configurable Mode

Scheduling for a partially re-configurable system is also described in detail in Part 2 Chap. 3, where reconfiguration of the VPs takes place independently during a time slice (other than the initial starting phase), with bitstreams for performing a particular subtask. The advantage of partially re-configurable mode is that unutilized slacks can be avoided and another subtask can be scheduled in individual VPs after completion of previous subtasks. However, it must be kept in mind that the VPs are reconfigured with the aid of ICAP ports, which are limited in a particular FPGA device. Hence, if there is only a single ICAP port, then only a single VP can be reconfigured at a single time and reconfiguration operation in another VP needs to wait if there is a simultaneous demand. Figure 5.3 is a pictorial representation of a task schedule for a partially re-configurable system that is described in detail in Part 2 Chap. 3. We take this for demonstration in the current chapter.

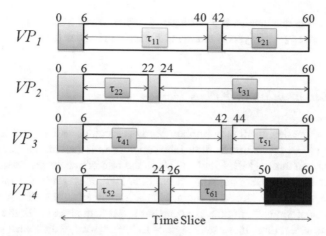

Fig. 5.3 Task schedule for a partially re-configurable system

5.3 Threat Model

Confidentiality attacks is associated with leaking secret information to adversaries via covert channels. This can take place via analysis of side channel parameters like timing, power, etc. This is generally caused by hardware trojan horses (HTHs) implanted either in the bitstreams or in the FPGA devices by adversaries during their design time.

For the present scenario that involves real time task schedules, an FPGA is shared both temporally and spatially by various users. As several tasks execute simultaneously, hence, it is difficult for adversaries to infer secret information about a specific task by monitoring the system power dissipation or leakage current of the system. However, adversaries may still use timing to reveal secret information as output for a particular task, as timing of each task is independent of others. Adversaries exploit the additional buffer time in task schedules that is allocated by designers to generate conditional delay for leakage of secret information.

The injected malware by adversaries may be either associated in the bitstreams or in the FPGA device, as described.

5.3.1 Vulnerability Present in Bitstreams

Malicious 3PIP vendors insert malicious codes during the hardware description language (HDL) phase of bitstream design that analyze the secret bits of the users and delays timing of output generation. Based on timing analysis over a range of operations, the adversary is able to get acquainted with the secret bits used by the user [LJM13, GSC17a]. As the malicious codes are quite small and dispersed in the sea of original codes, hence, it is difficult to detect. It has also been seen that even 100% code coverage analysis is unable to detect HTH codes [LJ01].

5.3.2 Vulnerability in FPGA Device

Malicious FPGA designers may implant hardware trojans either in the unused portions of the FPGA fabric during their fabrication or in the interconnects or buses or even in FPGA memory [MKN+16]. HTHs in the interconnects and buses can monitor the secret bits of users transmitted via them, while HTHs in the memory may monitor the secret data stored in them. HTHs implanted in the FPGA fabric may monitor the operations during their executions to get acquainted with the secret bits. Based on the secret data, these send signals to the payload, which varies the timing by conditionally delaying output generation, based on the secret key bits.

The structure of HTH that causes confidentiality attacks is as follows. A trigger module and a payload module are the key components of the HTH. The HTH trigger module can be either external or internal. For external trigger, reception of signals takes place via embedded antennas or sensors. Internal trigger is like a time bomb that gets activated by satisfaction of some pre-decided criterion set by adversaries like a certain number of operations or some specific combination of internal node values. Figure 5.4 depicts a time bomb based HTH trigger, where there is a counter c, which increments its count in each operation till it reaches a designated value v, set by the adversary. When c is equal to v, it sets the *Activating Signal*, which in turn activates the payload.

For getting acquainted with the secret bits, there is a key stealing module [LJM13]. The secret bits are stored in memory and generally kept unaltered for a series of operations. Considering an n bit secret data is loaded, the enable line is triggered to 1. Thus, secret bits that are stored in a set of D-Flip Flops (FFs) and latches as depicted in the normal scan chain of Fig. 5.5. The key idea for secret key bits extraction without hindering system performance is to reuse the D-FFs as a rotator, while values in

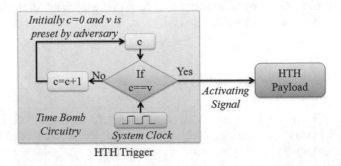

Fig. 5.4 Time bomb based HTH trigger

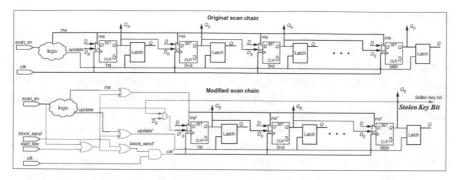

Fig. 5.5 Secret bit stealing circuitry [LJM13]

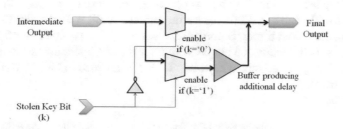

Fig. 5.6 HTH payload structure for leakage of secret bits based on timing

latches are to remain unaffected. This is facilitated by use of multiplexers before D-FFs and additional signals, as shown in the modified scan chain of Fig. 5.5. In this case, enable line is triggered only if a new secret key is loaded. This causes loading the D-FFs and associated latches with the new values. For other operations, i.e. when secret key is unaltered, then the enable line remains at 0. This causes working of the rotator, which involves propagation of secret key bits via the chain of D-FFs, without hampering the values stored in the latches. Thus, normal system performance takes place via the values of latches, but with each operation via the rotator, the adversary gets acquinted with one bit of secret key bit, as shown in the modified scan chain of Fig. 5.5.

When HTH Trigger is activated, as per the value of the stolen key bit, HTH payload varies a side channel parameter (which in the present case is timing) like delay or power to leak the stolen key bit to an adversary. Figure 5.6 depicts the payload operation, which is associated with a delay in output generation. Thus, at runtime, post Trojan activation, if value of stolen bit is 0, output arrives normally but if value of stolen bit is 1, an additional amount of delay is appended in output generation. Thus, the secret key is revealed to an adversary who monitors the output timings. Leakage of secret information for such scenarios will take place via a set of operations, which will be equivalent to the length of the secret information.

Demonstration

Let us consider a real time cryptographic task operation in a real time schedule, which uses a six bit secret key 101001. Assuming the affected task is T_{11} for the fully re-configurable system mode and τ_{61} for the partially reconfigurable system mode. The worst case execution time dedicated for T_{11} is 14 time units, while for τ_{61} is 24 time units. Let the procured IP deployed for the T_{11} task operation actually takes 10 time units for operation, while the procured IP deployed for τ_{61} take 20 time units for operation. The adversary steals the stolen key and tries to reveal the stolen key bits via 6 operations by varying the output timings. For this, in the first operation after HTH trigger, a delay of 2 time units is made, and thus, the output timing is at $t = 38$ and not $t = 36$ for T_{11} and for τ_{61}, output timing is $t = 48$ and not $t = 46$. In the second operation, as the stolen secret bit is 0, no additional delay is made and the output timing is $t = 36$ for T_{11} and $t = 46$ for τ_{61}. For third operation, output timing is $t = 38$ for T_{11} and $t = 48$ for τ_{61} as the stolen key bit is 1. For fourth and fifth operations, the output timing is normal, i.e. at $t = 36$ for T_{11} and $t = 46$ for τ_{61} as stolen key bit is 0. Finally, the output timing for the last operation is $t = 38$ for T_{11} and $t = 48$ for τ_{61}, as the stolen key bit is 1. This is depicted in Fig. 5.7 (only related VPs are shown).

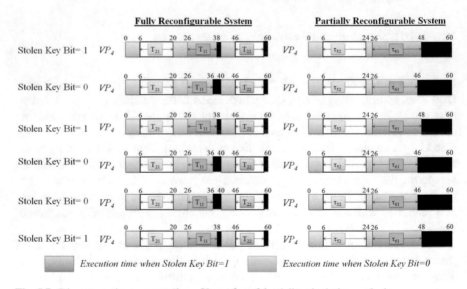

Fig. 5.7 Diagrammatic representation of loss of confidentiality via timing analysis

5.4 Self Aware Security to Bypass Passive Threats

5.4.1 Existing Strategies and Limitations

Existing mechanisms that ensure prevention from such threats involve obfuscation, algorithm balancing and masking [CB11, AZM10, JIP17]. Key based obfuscation techniques are quite common, where logic locking is performed during the IP design as a gate level design obfuscation technique to protect the outsourced IP designs from piracy and counterfeiting by adversaries in untrustworthy foundries. In such mechanisms, only if the correct key is provided at runtime, then only the circuit completes its functionality [CB11]. In algorithm balancing techniques, designers develop the algorithms in such a manner that the concerned IPs will thwart a difference in side channel parameters like delay and power, such that the internal structure of the IP is not revealed to an adversary [JIP17]. In masking techniques, additional logic is embedded in an IP design to mitigate difference in side channel parameters via which an adversary can infer the internal structure of an IP [AZM10].

However, all such techniques are associated with the design of a safe IP and facilitates protection from adversaries in the foundry that may implant HTHs in the layout during fabrication of the IP. In other words, we can call such techniques intrusive in nature, i.e. implantation of security elements is associated during design time and ensures protection from piracy and counterfeiting of the designs.

These do not address the scenario where the RIPs are procured from untrustworthy 3PIP vendors and the designer is unaware of the internal structure of the RIPs procured [LRYK14, RSK16]. Moreover, such RIPs may be packaged and be in an unreadable format to prevent piracy and code stealing. Hence, a non-intrusive mode of security is needed, where based on runtime performance, detection and mitigation of vulnerability can be ensured.

5.4.2 Security Mechanism

To mitigate revealing of secret information via timing analysis, it is necessary to deliver the output at a fixed pre-decided time instant. However, as different RIPs follow different algorithmic approach, hence, their execution times will be different and in a scenario that is devoid of any monitoring mechanism, output timing may vary. Hence, it is imperative to deploy a self aware agent (SAA) that will ensure output delivery at fixed time instant, irrespective of the time in which task execution completes. For the present case, it is worthy to deliver output at the end of the allotted execution time slot or at the end of the worst case execution time of a particular task. For this, additional wait time must be provided by the self aware module. For example, output of T_{11} for the fully re-configurable task schedule considered must be generated at time instant $t = 40$ and output for τ_{61} for the partially re-configurable task schedule must be generated at time instant $t = 50$, irrespective of task completion time.

The self aware module must not only align the timing of output generations of each task in the real time schedule, but must also detect such vulnerability and find out the affected resource or source of the vulnerability and take appropriate measures to mitigate such issues in future operations.

For detection of such vulnerability, the self aware module needs the aid of the task schedule generated offline. If the task completion time by a bitstream remains same or does not vary with time (which is less than the worst case execution time), then the scenario is normal. However, an increase in task completion time may either be associated with aging of the FPGA platform or due to HTH attack. Aging is a natural phenomenon associated with all hardware platforms [HLK+15]. However, aging before time may be due to counterfeit components or due to delay inducing HTHs. Such induced aging may cause a deadline miss and that is a case of active attacks [GSC17b, GSC19], which will be covered in the next chapter. Attacks due to HTH that will result in loss of confidentiality and will be associated with a scenario of delay and no delay or a scenario of more delay and less delay. Such conditions will indicate secret bits 1 and 0. The SAA on detecting such variation in task completion time will understand that the scenario is anomalous.

To prevent such an anomaly, the SAA must perform fault diagnosis to determine which resource, i.e. bitstream or the FPGA device is associated with a vulnerability. For this, another bitstream or RIP procured from another vendor is to be used for a future set of operations. If the problem persists, then vulnerability is associated with the FPGA platform and not the bitstream. However, no such vulnerability occurs, then the bitstream is affected. If the bitstream is affected, then the SAA must blacklist the vendor who supplied the bitstream and signal the FPGA controller so that no other bitstreams from that blacklisted vendor is used in future operations. Else raise an alarm, so that the system administrator replaces the affected FPGA resource.

5.4.3   Working of Self Aware Agent (SAA)

The SAA must be an external entity and not part of the FPGA device, i.e. the security logic of the SAA must not be embedded either in the programmable logic (PL) or processing system (PS) of the FPGA. This is of utmost importance as if the FPGA device is affected, then the self aware agent will also be functioning maliciously. Hence, logic or code of the SAA must be implanted in a daughter FPGA device that is placed outside the FPGA platform, as shown in Fig. 5.8. Task input and output takes place via the self aware agent. For working of the SAA, it must have access to the task schedule that is embedded in the FPGA memory and works based on the Observe-Decide-Act paradigm.

Observe Phase In the observe phase, the self aware agent notes the timing of the task input and task output for each task executed in its host FPGA platform. For this, it maintains a small memory unit where it stores for each task, the difference of actual task completion time and the worst case execution time allocated by the designer.

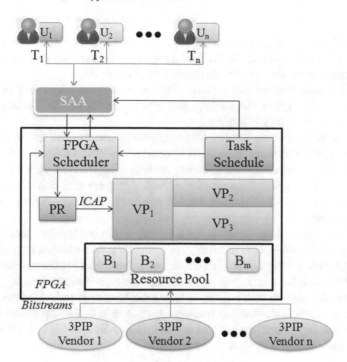

Fig. 5.8 System model with SAA

Decide Phase For a particular bitstream or RIP, the task completion time does not change in normal scenario. Hence, the normal scenario can be easily deciphered by the self aware agent, which initially finds the difference of the worst case execution time allotted by the designer and the present task completion time and compares this to the value stored in the memory for the particular task.

However, if at any stage, if it finds that the difference is not equivalent to the value stored in memory, then it will trigger the fault diagnosis phase.

In fault diagnosis, it checks whether the old difference value is greater or smaller than the new difference value. If the new difference value is lesser than the old difference value, then this can either take place due to aging or due to HTH attack. As no confirmation cannot be made in this stage, hence, it replaces the old difference value with the new difference value. In subsequent operations, if the new difference value keeps on decreasing than the difference value stored in its memory, then aging is confirmed for sure.

However, if in subsequent operations, the new difference value is more than the difference value stored in its memory, then this is an indication of HTH attack that is destroying confidentiality of the system by leaking secret bits to an adversary.

Act Phase

For Normal Scenario: Though task completion time takes place before the worst case execution time allotted, the self aware agent delays output generation till the current time is equivalent to the worst case execution time.

For Anomalous Scenario/when Fault Diagnosis is triggered: If aging is confirmed, then signal is raised for system administrator to replace it, else situation may arise where deadline miss may take place.

However, if confidentiality attack is confirmed, then the self aware agent will need to find out whether the bitstream is affected or the FPGA device is affected. For this, it will signal the controller to deploy another bitstream carrying out the same functionality but procured from a different vendor. If delay variation continues, then the FPGA device is affected and signal will be raised for its replacement, similar to the aging scenario.

However, if the FPGA device is not affected, or delay variation does not occur after bitstream is changed for the operation, then it is confirmed that the bitstream is affected. In such a scenario, the self aware agent will blacklist the vendor. For this to take place, one bit memory is to be allotted for each vendor. Initially, for each vendor, the memory bit will remain 1. When the vendor is to be blacklisted then, the memory bit dedicated for this vendor is to change to 0. The controller will only select bitstreams procured by vendors whose memory bit status is 1. On finding it 0, bitstreams associated with it will not be chosen.

5.4.4 Algorithm and Explanation of Proposed Mechanism

Algorithm 1 depicts the proposed mechanism, which is also represented in Fig. 5.9.

Let's consider t_{ipe} be the time taken for present execution for task i, t_{ine} denote time taken for normal execution of task i, t_{ios} represent output time of task i as per schedule, fd be a variable used during fault diagnosis for checking vulnerability due to FPGA, which is initially set to 0, x be a variable that denotes time out limit for checking bitstream vulnerability during fault diagnosis and c be a counter for checking vulnerability due to bitstream.

t_x is a temporary variable which is set with the value of t_{ipe}. Unless t_x is equivalent to t_{ios}, output generation is delayed by one time unit. This aids to align timing in all iterations and prevent leakage of secret information or confidentiality attack via timing analysis.

Now, t_{ipe} is equivalent ot t_{ine}, then the scenario is normal and no remedy is required. But if t_{ipe} is greater than t_{ine}, then there is an anomaly, which may be either due to aging or due to HTH attack. Inferring at this stage is not possible and is subject to more analysis. For further analysis, t_{ine} is updated with the value of t_{ipe}. An increase in t_{ipe} is not harmful, till it exceeds the allowable time limit in the schedule. If the limit is exceeded, then it is an active attack, which will be taken up in the next chapter.

However, if t_{ipe} is less than t_{ine}, then an attack to confidentiality is confirmed. This leads to triggering of fault diagnosis module. In fault diagnosis, it is to be

---

Algorithm 1: Proposed Mechanism

Input:

t_{ipe}: time taken for present execution for task i;

t_{ine}: time taken for normal execution of task i;

t_{ios}: Output time of task i as per schedule;

fd: Variable used during fault diagnosis for checking vulnerability due to FPGA, initially set to 0;

x: Time out limit for bitstream vulnerability observing during fault diagnosis;

c: Counter for checking vulnerability due to bitstream;

Output:

Normal Scenario: Output result;

Anomalous Scenario: Output FPGA affected or Bitstream Affected

for *Task i* **do**

    begin

        Set $t_{ipe} == t_x$; where t_x is a temporary variable;

        if $t_x == t_{ine}$ **then**

            Generate Output

        else

            $t_x == t_x + 1$

    if $t_{ipe} == t_{ine}$ **then**

        Normal Scenario

    else if $t_{ipe} > t_{ine}$ **then**

        Aging or Attack and hence, Update $t_{ine} with t_{ipe}$, i.e. $t_{ine} > t_{ipe}$;

    else if $t_{ipe} < t_{ine}$ **then**

        Attack Confirmed and hence, Trigger Fault Diagnosis;

if *Fault Diagnosis Triggered* **then**

    Set $fd = fd + 1$,

    if $fd > 1$ **then**

        FPGA is affected and generate Alarm;

    else

        Change bitstream from another vendor and perform task i;

        $c = c + 1$;

        if $c == x$ **then**

            Bitstream is affected and blacklist vendor from whom bitstream is procured;

            Reset fd to 0;

        else

            Increment c with each operation and check till timeout or $c == x$;

---

deciphered whether the FPGA is affected or the bitstream is affected. If fault diagnosis is triggered, variable fd is incremented. Initially, fd is 0 and on increment, $fd = 1$. Then, normal task execution is followed but with bitstream from another vendor. In addition to this, a counter is started and its count is incremented with each operation till it reaches the time out limit that is set by the designer, which in the present context, is represented by variable x. If within the time out limit, the same anomalous behavior is experienced, then fd becomes 2 and it can be confirmed that vulnerability exists with FPGA and alarm is generated. However, if no such anomaly is exhibited and

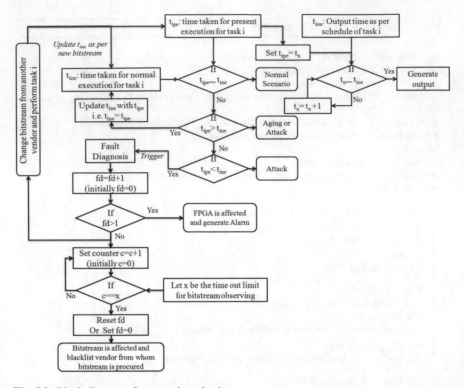

Fig. 5.9 Block diagram of proposed mechanism

the time out limit expires, then, it can be confirmed that vulnerability exists with the bitstream and the vendor from which the bitstream is procured is blacklisted. fd is reset to 0 to handle such anomaly in future.

5.4.5 Demonstration

For task T_{11} of the fully re-configurable schedule, the SAA on finding output at time instant either at $t = 38$ or $t = 36$, delays output generation till $t = 40$. For task τ_{61} of the partially re-configurable schedule, the SAA on finding task completion before the worst case time period, i.e. either at $t = 48$ or $t = 46$, prevents output generation till time reaches $t = 50$. This is depicted in Fig. 5.10. As, output generates at $t = 40$ for all operations of T_{11} and at $t = 50$ for all operations of τ_{61}, irrespective of the stolen key bit, hence, the adversary is unable to decipher the secret key bits via timing analysis. This is depicted diagrammatically in Fig. 5.10.

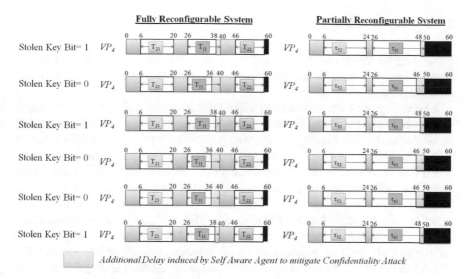

Fig. 5.10 Demonstration of the working of SAA

5.5 Experimentation and Results

5.5.1 Experimentation

Experimental Setup Simulation based experiments is performed for experimentation. 50 types of hardware task operations are considered in the task set for experimentation that ranges from standard ISCAS and ITC 99 benchmarks to cryptographic task operations of DES and AES crypto architectures. Bitstreams are generated in Xilinx Vivado platform. The SAA logic is implanted in a daughter FPGA, which monitors the performance of its host FPGA and performs related security functions. This is because if the SAA logic is implanted in a part of the host FPGA, then if the FPGA gets affected, then the SAA will cease to perform or perform maliciously.

Threat Generation For threats related to bitstreams, additional malicious logic or an HTH is implanted during the HDL design phase of bitstream generation.

For threat related to FPGAs, an additional small VP is generated, in addition to the normal VPs for task execution. In this additional VP, HTH logic is implanted.

The malicious logic or HTH comprises a five bit counter as its trigger. HTH payload is activated when the counter reaches 11111. As discussed previously, in addition to the payload, there exists an information stealing circuitry. The payload comprises of a conditional delay circuitry. The payload on getting activated, causes conditional delay based on the stolen secret bit.

For the present work, we depict key stealing for operations related to DES and AES crypto architectures that are considered for experimentation.

5.5.2 Result Analysis

Experimental Validation Threat on DES and AES crypto operations and related solution on application of our proposed mechanism is depicted in Fig. 5.11 and Fig. 5.12 respectively.

A secret key of 64 bits is associated with a DES crypto operation. For experimental validation, as per the offline schedule generated, time of output generation is at 33.5 ns. However, the DES operation completes before time and is able to deliver output at 30.2 ns. Thus, time from 30.2 to 33.5 ns can be considered as the buffer time, that is intentionally provided by the designer in case of some delay occurs during execution. During normal scenario or before HTH trigger, output takes place at 30.2 ns. Let HTH activation occur after 60 operations. Then, if the stolen key bit is '0', then output is delivered at 30.2 ns, but if the stolen key bit is '1', output is delivered at 32.2 ns. Thus, a 2 ns delay is generated, as per the stolen key bit. The adversary monitors such delay and is able to get acquainted with the secret key after 64 operations. But on application of our proposed mechanism, output generation is delayed and only takes place at 33.5 ns. This takes place whether HTH is activated or not. Hence, the adversary will not be able to decipher the secret key and the threat is bypassed. This is shown in Fig. 5.11.

AES crypto operation is associated with a 128 bit secret key. For our present experimentation, time for output generation of AES operation, as per the offline schedule generated is at 52ns. However, AES operation can complete before that time and output can be generated at 48.8 ns. Thus, buffer time exists from 48.8 to 52 ns. In normal scenario, all outputs are generated at 48.8 ns. However, when HTH triggers after 60 operations, then for stolen key bit '0', output generates at 48.8 ns, while for stolen key bit '1', output generates at 50.8 ns. An adversary monitors the

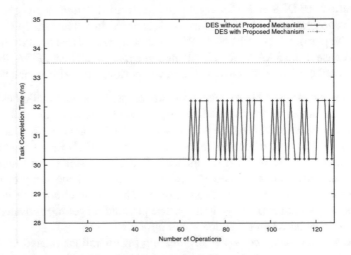

Fig. 5.11 Experimental validation on DES crypto operation

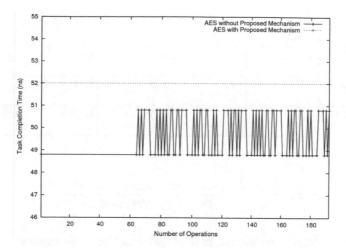

Fig. 5.12 Experimental validation on AES crypto operation

timing and is able to know the secret key, after monitoring 128 successive operations. However, on application of our proposed mechanism, output is generated at 52 ns irrespective of whether HTH gets activated or not and also irrespective of the stolen key bit. Hence, the threat is bypassed. This is depicted in Fig. 5.12.

Overhead Analysis For experimental validation, we increase the number of partitions of an FPGA from 1 to 16, i.e. facilitate number of parallel task processing from 1 to 16 and analyze the overhead incurred by the SAA with respect to area and power. The task schedules are accordingly modified, with increase in the number of task handling.

With increase in the number of VPs, which indirectly represents increase in number of task handling, the SAA has to carry out additional functionalities of monitoring each and every task. Hence, it's overhead with respect to area and power increases accordingly.

Area overhead is defined as the percentage ratio of amount of resources of security elements to the amount of resources of the original system. Resources in the present case is the number of LUTs and slices, as the system under consideration is an FPGA. This can be represented by the following equation.

$$Area\ Overhead = \frac{Resources\ of\ Security\ Elements}{Resources\ of\ Original\ System} * 100\% \qquad (5.1)$$

Similarly, power overhead is defined as the ratio of increment in power due to addition of security elements to the original power incurred by the system, expressed as a percentage. Power is measured in mW. This is represented by the following equation.

Table 5.1 Overhead analysis of SAA with increase in the number of task handling

	Number of tasks (2^k)				
Overhead of SAA	k = 0	k = 1	k = 2	k = 3	k = 4
% increase in area	1.41	1.61	2.15	3.51	5.61
% increase in power	0.50	0.59	0.79	1.13	2.59

$$Power\ Overhead = \frac{Additional\ Power\ due\ to\ Security\ Elements}{Power\ of\ Original\ System} * 100\%$$

$$(5.2)$$

Table 5.1 presents the increase in overhead with respect to area and power for a SAA, with increase in the number of tasks for an FPGA. As evident, with increase in number of tasks, overhead in area and power increases. This is due to the fact that additional monitoring and decision making must be performed by the SAA. However, the overhead is quite nominal and for a system with 16 VPs, overhead in area is less than 6% and overhead in power is less than 4%.

5.6 Conclusion

In this chapter, we analyze how confidentiality attacks may take place on real time tasks, associated with secret information, in a task schedule. We propose development of a SAA that monitors such activity and bypasses such threats by maintaining pre-decided output timings of all task operations, as per the offline generated task schedule. The SAA on detecting anomalous activity, performs fault diagnosis to decipher whether the bitstream is affected or the FPGA is affected. If the FPGA is affected, it generates an alarm for its replacement. Else, if the bitstream is affected, the 3PIP vendor from whom the bitstream is procured is blacklisted. Experimental validation is performed with DES and AES crypto operations to demonstrate the threat and related mitigation technique. Overhead analysis of the SAA with respect to area and power is performed with increase in the number of VPs of the FPGA, i.e. increase in the number of tasks. Nominal overhead of the SAA depicts its applicability for practical applications.

References

[AARP10]   J. Aarestad, D. Acharyya, R. Rad, J. Plusquellic, Detecting trojans through leakage current analysis using multiple supply pad I_{DDQ}s. IEEE Trans. Inf. Forensics Secur. **5**(4), 893–904 (2010)

[AZM10]   A. Askarov, D. Zhang, A.C. Myers, Predictive black-box mitigation of timing channels, in *Proceedings of the 17th ACM Conference on Computer and Communications Security, CCS '10* (2010), pp. 297–307

[CB11]   R.S. Chakraborty, S. Bhunia, Security against hardware trojan attacks using key-based design obfuscation. J. Electron. Test. **27**(6), 767–785 (2011)

[GSC15]   K. Guha, D. Saha, A. Chakrabarti, RTNA: securing SOC architectures from confidentiality attacks at runtime using ART1 neural networks, in *2015 19th International Symposium on VLSI Design and Test* (2015), pp. 1–6

[GSC17a]   K. Guha, D. Saha, A. Chakrabarti, Real-time soc security against passive threats using crypsis behavior of geckos. J. Emerg. Technol. Comput. Syst. **13**(3), 41:1–41:26 (2017)

[GSC17b]   K. Guha, D. Saha, A. Chakrabarti, Self aware SoC security to counteract delay inducing hardware trojans at runtime, in *2017 30th International Conference on VLSI Design and 2017 16th International Conference on Embedded Systems (VLSID)* (2017), pp. 417–422

[GSC18]   K. Guha, S. Saha, A. Chakrabarti, Shirt (self healing intelligent real time) scheduling for secure embedded task processing, in *2018 31st International Conference on VLSI Design and 2018 17th International Conference on Embedded Systems (VLSID)* (2018), pp. 463–464

[GSC19]   K. Guha, D. Saha, A. Chakrabarti, Stigmergy-based security for SoC operations from runtime performance degradation of SoC components. ACM Trans. Embed. Comput. Syst. **18**(2), 14:1–14:26 (2019)

[Guh20]   A. Majumder, D. Saha, A. Chakrabarti, K. Guha, Dynamic power-aware scheduling of real-time tasks for FPGA-based cyber physical systems against power draining hardware trojan attacks. J. Supercomput. **76**(11), 8972–9009 (2020)

[HLK+15]   K. Huang, Y. Liu, N. Korolija, J.M. Carulli, Y. Makris, Recycled IC detection based on statistical methods. IEEE Trans. Comput. Aided Des. Integr. Circuits Syst. **34**(6), 947–960 (2015)

[JIP17]   D. Jayasinghe, A. Ignjatovic, S. Parameswaran, NORA: algorithmic balancing without pre-charge to thwart power analysis attacks, in *2017 30th International Conference on VLSI Design and 2017 16th International Conference on Embedded Systems (VLSID)* (2017), pp. 167–172

[LJ01]   C. Liu, J. Jou, Efficient coverage analysis metric for HDL design validation. IEE Proc. Comput. Digit. Tech. **148**(1), 1–6 (2001)

[LJM13]   Y. Liu, Y. Jin, Y. Makris, Hardware trojans in wireless cryptographic ICs: silicon demonstration & detection method evaluation, in *Proceedings of the International Conference on Computer-Aided Design, ICCAD '13* (2013), pp. 399–404

[LRYK14]   C. Liu, J. Rajendran, C. Yang, R. Karri, Shielding heterogeneous MPSoCs from untrustworthy 3PIPs through security-driven task scheduling. IEEE Trans. Emerg. Top. Comput. **2**(4), 461–472 (2014)

[MKN+16]   S. Mal-Sarkar, R. Karam, S. Narasimhan, A. Ghosh, A. Krishna, S. Bhunia, Design and validation for FPGA trust under hardware trojan attacks. IEEE Trans. Multi-Scale Comput. Syst. **2**(3), 186–198 (2016)

[NDC+13]   S. Narasimhan, D. Du, R.S. Chakraborty, S. Paul, F.G. Wolff, C.A. Papachristou, K. Roy, S. Bhunia, Hardware trojan detection by multiple-parameter side-channel analysis. IEEE Trans. Comput. **62**(11), 2183–2195 (2013)

[RB18]   Luis Ribeiro, Mats Björkman, Transitioning from standard automation solutions to
            cyber-physical production systems: an assessment of critical conceptual and technical
            challenges. IEEE Syst. J. **12**(4), 3816–3827 (2018)
[RSK16]   J.J. Rajendran, O. Sinanoglu, R. Karri, Building trustworthy systems using untrusted
            components: a high-level synthesis approach. IEEE Trans. Very Large Scale Integr.
            (VLSI) Syst. **24**(9), 2946–2959 (2016)

Chapter 6
Counteracting Active Attacks

6.1 Introduction

Active threats are associated with attacks that cause direct damage to a system, eventually jeopardizing it. This comprise either generation of erroneous results [MWP+09, GSC19a] or preventing result generation within deadline [GSC17, GSC19b]. The former is an attack to system integrity, while the latter is an issue related to system availability.

Real time FPGA based embedded systems are designed for mission critical applications like automated surveillance, medication, communication, navigation, automobile braking, process control, etc. [HKM+14]. These are associated with execution of a number of tasks with strict deadlines. Hence, they are ordered in schedules [GSC18]. Active attacks may disrupt individual task operations of such schedules or may even jeopardize the entire schedule, resulting in fatal consequences.

As discussed in previous chapters, task operations in such FPGA based systems are associated with procuring of various bitstreams or reconfigurable intellectual properties (RIPs) from various vendors and scheduling them in the FPGA at related time frames, as decided in the offline generated real time task schedule. However, trustworthiness of the third-party vendors is a concern. HTHs may be implanted in the RIPs by untrusted third-party vendors [LRYK14, RSK16].

Moreover, a globalization procedure is followed in the development of FPGAs that are procured and implanted in systems by the system designers [BT18]. Hence, HTHs may also be implanted by adversaries in the empty spaces of a layout during their fabrication [XFT14]. Even designers may be supplied with counterfeit FPGAs [CD07].

HTHs may induce sudden unexpected delays at runtime, preventing task completion within the pre-decided time frames of the related schedule [GSC18]. HTHs may also change functional operations leading to erroneous results [MWP+09, AAS14, GSC19a]. Use of counterfeit FPGAs are associated with degradation of performance at runtime, which may also cause deadline miss for the real time task operations [GZFT14]. Thus, system integrity and availability are at stake.

© The Author(s), under exclusive license to Springer Nature Switzerland AG 2021   111
K. Guha et al., *Self Aware Security for Real Time Task Schedules in Reconfigurable Hardware Platforms*, https://doi.org/10.1007/978-3-030-79701-0_6

Prevailing hardware security strategies focus on detection of HTH infected RIPs [XZT13, NDC+13] or counterfeit components [GDT14, ZWB15]. However, such compromised components will continue to function until and unless these are replaced [CMSW14]. Hence, recovery based strategies is not only important but essential. Strategies promising recovery at runtime are still in their infancy. Existing redundancy based runtime security strategies are essentially applicable for detection of erroneous result generation [MWP+09]. In this, the same task is executed simultaneously in multiple IP cores, procured from different sources. After task completion in all IP cores, fault diagnosis is applied to select the correct result, via a comparison or polling methodology. Such an approach endures a high overhead in area and power, with low throughput. Moreover, simultaneous focus to mitigate the issue of deadline miss due to delay inducing HTHs or counterfeits is also not underlined in such works. In addition to these, the works demonstrate security for standalone working environments and not for real time task schedules.

In the present chapter, we analyze how vulnerability of hardware may cause active attacks to real time task schedules for systems with a single or multiple FPGA platforms. This may either affect the integrity or availability of the system. It is difficult to analyze the internal structure of the FPGAs or the RIPs that are procured from third party vendors, as these are in unreadable and packaged format. To facilitate non-intrusive security in such a scenario, low overhead self aware agents (SAAs) must be designed and deployed that possess the ability to sense the environment and organize the behavior at runtime to mitigate active attacks. Like the previous chapters, the SAAs also work based on the Observe-Decide-Act (ODA) paradigm [SDG+15]. For a multi-tasking system, the SAAs communicate among themselves and assure security via a decentralized control.

The present chapter is organized as follows. Section 6.2 describes the system model, while the threat model is discussed in Sect. 6.3. Section 6.4 presents the security mechanisms for single and multiple FPGA platforms. Experimentation and results are discussed in Sect. 6.5 and the chapter concludes in Sect. 6.6.

6.2 System Model

We consider a multi user system with either a single FPGA resource or a number of FPGA resources. Users send their tasks, which are collected by the scheduler and allocated to the FPGA/FPGAs as per the real time schedules designed offline.

6.2.1 Single FPGA Based System

For a single FPGA based system, the scheduler may be present directly in the FPGA giving signals to the partial reconfiguration (PR) module to reconfigure the virtual portions (VPs) of the FPGA fabric at appropriate time slots with relevant bitstreams

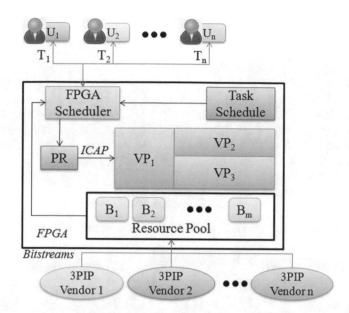

Fig. 6.1 System model for single FPGA based platform

present in the FPGA memory, via the Internal Control Access Port (ICAP) as shown diagrammatically in Fig. 6.1. The bitstreams are procured from various third party IP (3PIP) vendors and stored in the resource pool.

6.2.2 Multi FPGA Based System

However, for multi FPGA systems, control may either be centralized or decentralized.

For centralized control, there must be a global scheduler and a local scheduler in each FPGA. Pre-decided task schedules are stored centrally and are accessible by the global FPGA scheduler. The global FPGA scheduler collects user tasks and based on the pre-decided task schedule, tasks are sent for execution to the respective local FPGAs at appropriate time instants. In such an architecture, the procured bitstreams from 3PIP vendors need not to be stored locally in each FPGA, and can be stored in a central resource pool. On getting tasks from the global scheduler, the local scheduler of each FPGA sends signals to its respective PR module to configure the related bitstream received from the resource pool on the appropriate VP via the ICAP port. On task completion, the result is sent back to the global scheduler, which delivers it to the respective user. A diagrammatic representation of the system architecture is shown in Fig. 6.2.

For illustrating decentralized control systems, let us consider the diagrammatic representation of Fig. 6.3. Each node comprises of a single FPGA that caters to a

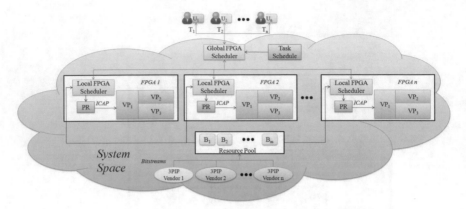

Fig. 6.2 System model for multi FPGA platform with centralized control

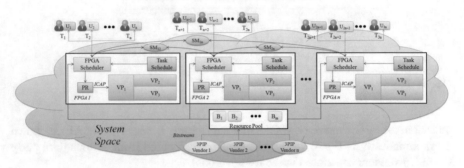

Fig. 6.3 System model for multi FPGA platform with decentralized control

number of users. Dedicated task schedules of the FPGAs that are designed offline, prior to deployment, are stored internally in each FPGA node. The FPGAs may allot tasks as per the schedules for execution in the respective VPs of the FPGAs with the aid of bitstreams procured from 3PIP vendors, via the PR modules and the ICAP port. The bitstreams may either be stored internally in each FPGA memory or may be centrally stored in a central resource pool as depicted in Fig. 6.3. Communication among the FPGAs take place via shared memories (SMs).

Modes of Operation The FPGAs may either perform in a fully re-configurable mode or in a partially re-configurable mode. Details of scheduling for either mode are discussed in previous chapters. The scheduling is performed offline that determines which task is to be scheduled at what time, in which FPGA VP. Initially, the work share of tasks for a time slice is determined, whose length is equivalent to the minimum deadline among all tasks in the task set. In the subsequent time slices, residual task shares are completed. Bitstreams or reconfigurable IPs (RIPs) from different third party IP vendors are used to execute the tasks/subtasks in the FPGA platform/s.

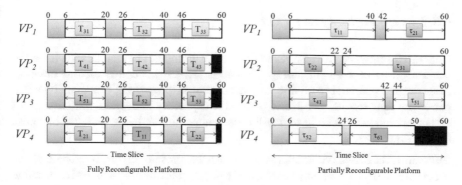

Fig. 6.4 Real time task schedules for fully and partially reconfigurable FPGAs

For fully reconfigurable mode of operation, reconfiguration of all VPs in an FPGA platform take place simultaneously in a time slice. While for a partially reconfigurable mode of operation, reconfiguration of VPs take place independently during a time slice (except for initial reconfiguration). Details of scheduling for fully reconfigurable mode of operation and partially reconfigurable mode of operation is discussed in Part 2 Chap. 3 respectively. Figure 6.4 depicts real time task schedules for fully and partially reconfigurable modes of operation, which are described in detail in Part 2 Chap. 3. These we use in the present chapter for demonstration.

6.3   Threat Scenario

In the current chapter, we consider active attacks that can hamper the basic security primitives, integrity and availability of the system. The former is associated with generation of erroneous results, while the latter causes denial of service (DoS), which prevents real time tasks to complete their execution within their designated time frame. Sources of vulnerability are described below.

6.3.1   Vulnerability in RIPs/Bitstreams

Malicious 3PIP vendors that design the bitstreams may insert malicious codes in the bitstreams or RIPs during the hardware description language (HDL) phase of bitstream generation [LRYK14, LRYK13]. Even 100% code coverage analysis is unable to detect them [LJ01], as discussed before.

6.3.2   Vulnerability in FPGAs

Implanted malware Adversaries in foundry may implant malicious circuits in the unused spaces of a layout during fabrication [XFT14]. Such malicious circuitry possesses the potential to additionally execute or skip essential functionalities that generate wrong results. These also posses the capability to degrade performance of system operations, which in turn will cause unintentional delays and deadline miss for real time tasks.

Interconnects/Buses Vulnerability may also be present in the interconnects or buses [MKN+16]. These may cause a delay in data transmission and lead to deadline miss, affecting real time task operations.

Counterfeit components As discussed in previous chapters, an FPGA comprises of several components that are procured from third party sources and integrated. However, if the components are counterfeit then they will exhibit a degraded performance at runtime and might also cease to operate before their lifetime ends [GZFT14, BT18]. Finite delays will affect system availability, while infinite delays and stoppage of operations additionally impact system integrity.

The malicious codes in the bitstreams, malicious circuitry in the FPGA fabric or the malicious implants in the interconnects can be collectively termed as hardware trojan horses (HTHs). As discussed before (in previous chapters), the HTHs comprise a trigger and a payload [BHBN14, TK10]. The trigger can be either internal or external. Internal trigger may be like a time bomb that activates after a certain pre-decided number of operations or when a rare combination of node values actuates, as set by the adversary. External trigger receives signals from an adversary via an antenna or sensors.

The payload encapsulates the malicious operation, which in the present case may either affect integrity or availability of the system. System integrity is affected when unintentional additional functional operations are performed or essential functionalities are skipped. System availability is affected when the payload comprises delay inducing elements. Delay induced can be either finite or infinite. Finite delays is caused by elements like buffers or complementing logic gate operations. However, if the payload comprises of a loop architecture, then the delay will be infinite in nature. Figure 6.5 provides a diagrammatic representation of such an HTH with a time bomb-based trigger and payload that affects system integrity or availability.

Other than HTHs, even counterfeit components degrade system performance at runtime [CD07, KPK08, BT18]. However, during testing, no anomaly is detected in the operations of counterfeits. Such degraded performance will cause unintentional delays in operations and prevent real time tasks to miss their deadlines. For example, if the dynamic clock management (DCM) component of the FPGA platform is counterfeit, then it will not be able to upgrade the operations to the appropriate frequency causing delay in operations and will prevent real time tasks to complete operations within their deadlines. Moreover, vulnerability in the partial reconfiguration (PR) module or the internal configuration access port (ICAP) may prevent proper reconfiguration of VPs at appropriate time instants and a delay in reconfiguration of the

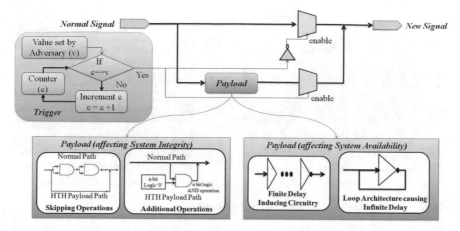

Fig. 6.5 HTH structure

VPs will prevent task completion at appropriate times. Moreover, if the input/output (I/O) port is affected, then delay may take place in task acceptance from users or task delivery to users. In either case, such delay may cause deadline miss. Even vulnerability in FPGA buses that are associated with transfer of data may prevent task completion within deadline. Moreover, if the SRAM memory of an FPGA is counterfeit, then delay will occur in memory access, which in turn may lead to delay in task execution causing deadline miss.

Demonstration for Real Time Task Schedules

Vulnerability in Bitstreams Initially, we demonstrate vulnerability associated with bitstreams. Let bitstream associated with T_{32} and τ_{22} of real time schedules depicted in Fig. 6.4 be associated with a vulnerability.

If system integrity is at stake, then output of T_{32} and τ_{22} will be generated at appropriate time instants but the results will be anomalous. Without any monitoring mechanism, the anomaly will go undetected. This is depicted in Fig. 6.6.

Figure 6.7 shows effect of delay associated with bitstreams for a fully reconfigurable platform and a partially reconfigurable platform, which affects system availability. Thus, T_{32} and τ_{22} will not be able to complete within time. However, as per the original schedules, reconfiguration will take place at pre-decided time instances. Moreover, if the tasks are periodic in nature, then malfunction will take place in each periodic execution, unless detection and mitigation is carried out.

Vulnerability in FPGA For demonstrating vulnerability in FPGA, we consider an HTH that gets activated at time instant $t = 30$ and causes malfunction. Let's consider that the malfunction be temporary in nature.

For integrity attacks, let some essential logic operations are skipped. Then, all task operations that are executing in that FPGA platform at that time instant will generate wrong results, as depicted in Fig. 6.8.

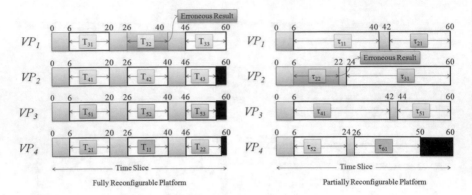

Fig. 6.6 Integrity attack due to vulnerability in bitstreams

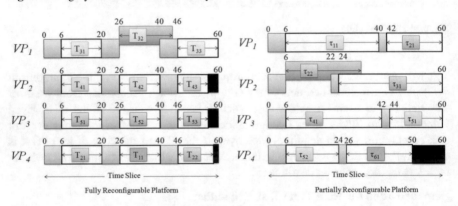

Fig. 6.7 Availability attack due to vulnerability in bitstreams

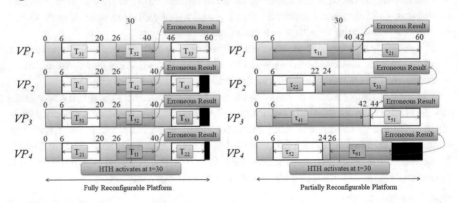

Fig. 6.8 Integrity attack due to vulnerability in FPGA

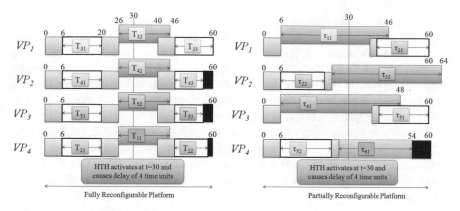

Fig. 6.9 Availability attack due to vulnerability in FPGA

For availability attacks, let the HTH causes unintentional delay of 4 time units. Hence, all tasks that are executing at that time instant will endure an additional delay. However, as slack space is available after task τ_{61}, hence, effect of unintentional delay is not effective for τ_{61}. For all others, they will not be able to complete within their allocated time interval. This is shown in Fig. 6.9.

Note It may so occur that only a certain portion of the FPGA fabric is associated with an error and not the entire FPGA. In such a case, task executions associated with only the FPGA VP that encompasses the faulty region will get affected. The scenario will become equivalent to vulnerability in bitstreams.

6.4 Redundancy Based Mechanism and Application to Current Context

Existing works depict how redundancy can be used to mitigate the issue of erroneous result generation or integrity attacks [MWP+09, AAS14]. In this approach, multiple IPs are procured from different vendors and the same task is simultaneously executed in them, followed by result checking or majority polling to select the correct result. At least three redundant executions or triple modular redundancy is used to detect the threat scenario and ensure correct output generation. Such a mechanism is even followed in some recent works for ensuring criticality based reliability for real time mixed critical task schedules [GMSC18, GMSC19].

Though the existing literature exhibit how redundancy can be applied for mitigating integrity attacks or counteracting erroneous result generation, however, no explicit application of the approach is present for counteracting availability attacks.

The underlying principle is that the probability of existence of HTHs having similar architecture, especially the trigger time, in resources procured from different vendors is negligible [RSK16, CMSW14]. For preventing integrity attacks, which not only involve correct output generation but also associates with detection of malfunctioning element, triple modular redundancy is essential. As for a dual redundant system, if results mismatch, it can be understood that one resource is functioning anomalously, but the malfunctioning resource cannot be identified. Hence, the third redundant operation is essential to identify the malfunctioning resource. The one whose result does not match the result of the third execution, is functioning anomalously and the majority result is generated as output.

For mitigating availability attacks via redundancy-based approach, triple modular redundancy is not needed. Double modular redundancy or executing a task in two resources procured from two different vendors is sufficient to mitigate threats related to availability. At least one result will generate within time, which can be delivered to the user.

6.4.1 Application for Single FPGA Based Platform

For a single FPGA based platform, only attacks related to bitstreams procured from untrustworthy 3PIP vendors is possible to mitigate. For issues related to integrity, it is neither possible to detect, nor mitigate attacks related to FPGA platform. For issue related to availability, though attacks on the FPGA platform is possible to detect and stop their operations, but mitigating them is not possible as there is no alternate FPGA platform where the task can be executed.

Fully Reconfigurable System Flexibility is limited for a fully reconfigurable system. Three instances of each subtask needs to be executed in the same time slice to mitigate integrity attacks. While for mitigating availability attacks, two instances of each subtask need to be executed in the same time slice.

If unutilized space or slacks are available, then no frequency increment is required. Else frequency of operation must be incremented to ensure task completion within the required time frame. The amount of frequency increment in each VP can be found out in the following manner. Considering time for full reconfiguration is t_{fr}, summation of total worst case execution time for all subtasks is t_e, total length of time slice is t, the total number of subtasks that needs to be executed is y. Then for the redundancy based approach, the total number of subtask instances that needs to be executed is $3y$ for mitigating integrity attacks and $2y$ for mitigating availability attacks. Hence, $3y$ and $2y$ number of full reconfigurations are required for the two scenarios. Thus, the amount of free time or time available (t_a) in a time slice available for execution is $t_a = t - (t_{fr} * 3y)$ for mitigating integrity attacks and $t_a = t - (t_{fr} * 2y)$ for mitigating availability attacks. If $t_a > 3 * t_e$ for mitigating integrity attacks and $t_a > 2 * t_e$ for mitigating availability attacks, then no enhancement of frequency is required. Else frequency enhancement is required by a factor of $t_a/3 * t_e$ for mitigating integrity attacks and $t_a/2 * t_e$ for mitigating availability attacks. Then as per the scheduling mechanism, the subtasks are to be scheduled for operation within the time slice.

Demonstration

Considering the fully reconfigurable schedule of Fig. 6.4, three task instances and two task instances of each subtask must be executed for mitigating integrity and availability attacks respectively.

In general, it is considered that $t_{fr} = 6$ time units, $y = 3$, $t = 60$ time units. In such a scenario, for mitigating integrity attacks, available time $t_a = 6$ time units. As at least 9 instances of subtask operations are required during available time and each task instance needs at least 1 time unit to operate, hence, this will not be possible. Hence, to demonstrate this particular case only, we consider $t_{fr} = 3$ time units, then available time $t_a = 33$ time units and $t_e = 126$ time units. Thus, each execution phase must be of 3 time units and frequency increment required is 5 times the normal frequency of operations. At the end, six time units remain as slack. Similarly, the trend continues for other VPs also, as depicted in Fig. 6.10a. As evident from the diagrammatic illustration, at least bitstreams from three vendors are required to mitigate integrity attacks.

Considering $t_{fr} = 6$ time units, $y = 3$, $t = 60$ time units, then for mitigating availability attacks, in a VP, at least 6 subtask instances must be executed. Then for VP_1, available time $t_a = 24$ time units and $t_e = 84$ and frequency increment required is 4 times the normal frequency of operation. The trend continues for other VPs also. The new schedule is as depicted in Fig. 6.10b. As evident from the diagrammatic illustration, at least bitstreams from two vendors are required to mitigate availability attacks. Different bitstreams are used for execution of the different task instances. If one is associated with a vulnerability, then the other will definitely complete within time and results of task instances will be generated safely before completion of the time slice.

However, the following problems will be associated:

(i) Several slacks present that is basically unutilized.

(ii) Huge increment in frequency of operations is needed (Sometimes this may exceed the maximum allowable frequency of operation).

(iii) Huge overhead in power.

Partially Reconfigurable System Other than the initial full reconfiguration, only partial reconfigurations are associated with the individual VPs. Similar to the previous one, increment in frequency for a VP can be calculated as follows. Assuming t_{fr} and t_{pr} are the time required for full and partial reconfigurations respectively, t_e as the summation of total worst case execution time for all subtasks and t the total length of time slice and y being the total number of subtasks that have to be executed.

Then like the previous scenario, for present redundancy based approach also, three instances for each subtask are required for execution to mitigate integrity attacks and two instances for each subtask are required to mitigate availability attacks. Thus, $3y$ and $2y$ are the total number of subtask instances that have to be executed to mitigate integrity and availability attacks respectively. Hence, available time can be calculated as $t_a = t - (t_{fr} + 3 * t_{pr} - t_{pr})$ for mitigating integrity attacks and $t_a = t - (t_{fr} + 2 * t_{pr} - t_{pr})$ for mitigating availability attacks. If $t_a > 2 * t_e$ then no enhancement of frequency is required, else frequency enhancement is required by a factor of

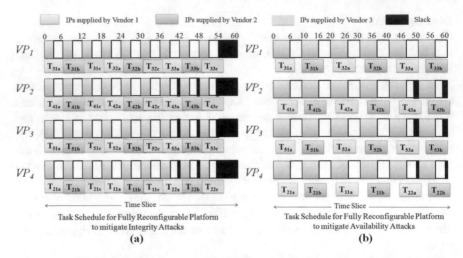

Fig. 6.10 Redundancy based task schedule for fully reconfigurable system to **a** mitigate integrity attacks **b** mitigate availability attacks

$t_a/(2 * t_e)$. Then as per the scheduling mechanism, the subtasks are to be scheduled within the time slice for execution.

Demonstration

For demonstration, let us consider the task schedule of partially reconfigurable platform of Fig. 6.4, where $t_{fr} = 6$ time units, $t_{pr} = 2$ time units, $y = 2$, $t = 60$ time units for VP_1.

Then for mitigating integrity attacks, 6 subtask instances has to be executed. Available time $t_a = 48$ time units and $t_e = 156$ and frequency increment required is 4 times the normal frequency of operation. Even 4 times frequency increment is required for VP_2 and VP_3, while 3 times frequency increment is required for VP_4, as depicted in Fig. 6.11a. As evident from the diagrammatic illustration, at least bitstreams from three vendors are required to mitigate integrity attacks.

Then for mitigating availability attacks, 4 subtask instances has to be executed. Available time $t_a = 48$ time units and $t_e = 104$ and frequency increment required is 3 times the normal frequency of operation. For VP_2 and VP_3, frequency increment is also 2 times, while for VP_4, frequency increment needed is 2 times. The new schedule is as depicted in Fig. 6.11b. Different bitstreams are used for execution of the different task instances. If one is associated with a vulnerability, then the other will definitely complete within time and results of task instances will be generated safely before completion of the time slice. As evident from the diagrammatic illustration, at least bitstreams from two vendors are required to mitigate availability attacks.

Though unutilized slack space is reduced, however, area and power overhead is severe.

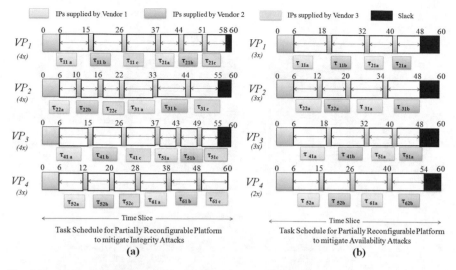

Fig. 6.11 Redundancy based task schedule for partially reconfigurable system to **a** mitigate integrity attacks **b** mitigate availability attacks

6.4.2 Application for Multi FPGA Based Platform

When not only security from vulnerability of bitstreams is required, but also security from vulnerability in FPGAs is also needed, then a switch to multi FPGA platform is required.

For both fully and partially reconfigurable systems, concurrent execution of subtasks in multi FPGA platforms are required. This is applicable for both the centralized and distributed system model, though the mechanism of control may vary. For mitigating integrity attacks, redundant task execution in three FPGA platforms are required, while for mitigating availability attacks, redundant task execution in two FPGA platforms are required.

For the centralized system, the global FPGA scheduler or controller collects user tasks and distributes them to the respective FPGAs. Majority polling of the results is performed to mitigate integrity attacks in the same global controller, before generating results to user. While for mitigating availability attacks, the global controller outputs the results that comes within time or before deadline.

However, for the distributed system model, the tasks and results are interchanged among the FPGAs via the shared memories. The controller that communicates with the user, is responsible to perform the majority polling operation before output generation for mitigating integrity attacks. While for mitigating availability attacks, the FPGA controller that communicates directly with the user outputs the results that arrives before deadline.

Demonstration We consider the fully reconfigurable system depicted in Fig. 6.4 for demonstration. Three FPGAs are used, i.e. $FPGA_1$, $FPGA_2$ and $FPGA_3$ for

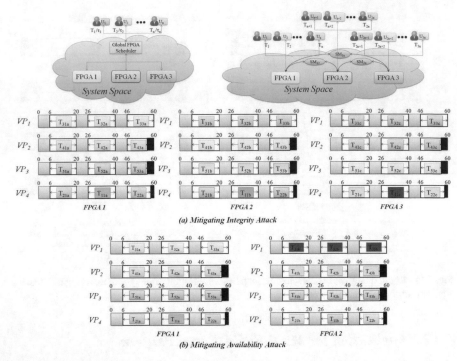

Fig. 6.12 Redundancy based task schedule for fully reconfigurable system to **a** mitigate integrity attacks **b** mitigate availability attacks

mitigation of integrity attacks, while two FPGAs, $FPGA_1$ and $FPGA_2$ are required for mitigating availability attacks.

For fully reconfigurable system, Fig. 6.12a depicts mitigation of integrity attacks, while Fig. 6.12b shows mitigation of availability attacks.

For partially reconfigurable system, Fig. 6.13a shows mitigation of integrity attacks, while Fig. 6.13b depicts mitigation of availability attacks.

6.5 Self Aware Mechanism

The problems of redundancy based approach can be mitigated by switching to a self aware mechanism. In a self aware approach, low overhead self aware agents are deployed that does not interfere with the working of the procured resources but work based on an observe-decide-act (ODA) paradigm. In the Observe Phase, the agents monitor the status. Based on the status, it deciphers the course of future action in the Decide Phase. Finally, in the Act phase, it either takes no action if the scenario is normal or if a threat occurs, takes necessary action to mitigate or bypass the threat.

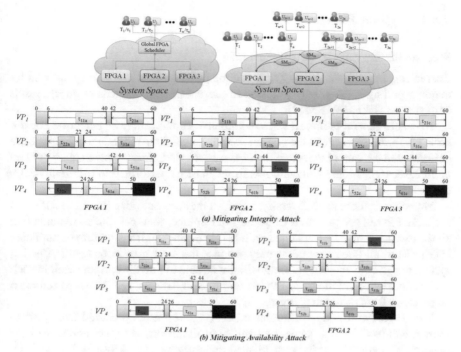

Fig. 6.13 Redundancy based task schedule for partially reconfigurable system to **a** mitigate integrity attacks **b** mitigate availability attacks

Ideal is the scenario where each IP for executing a task/subtask in a time slice of a schedule is procured from a different vendor. However, this will increase cost tremendously. This can be controlled if multiple IPs for different task/subtasks operation are procured from a particular vendor. However, diversity is important. Without diversity, i.e. the scenario where all IPs for executing the tasks/subtasks are procured from a single vendor, possibility remains that the same HTH is implanted in all the supplied IPs. Such attack scenario is also termed as duplicate HTH attack [GSC19b]. Then all the IPs will function maliciously at the same time. Detecting the anomaly will be difficult and, in some cases, it can be erroneously concluded that the FPGA device is faulty, which is not the case.

The security mechanism is basically an offline-online approach. In the offline phase, a periodic task schedule is generated based on the security needs of the tasks. In the online phase, the agents monitor the activities and ensure security.

6.5.1   Offline Phase

Security Strategy as per Attack Scenario

Preventing Integrity Attacks Triple redundancy in a task execution, followed by majority polling is essentially followed to prevent integrity attacks or ensure result correctness [MWP+09, BM13].

To handle general faults, all three redundant operations may be performed on the same platform. However, in the present scenario, if all three redundant operations are performed in the same FPGA and the FPGA is affected, then the mechanism will be ineffective. Hence, the redundant operations must be performed in different FPGA platforms.

Moreover, if there is no diversity, i.e. if all the resources like the three RIPs and the three FPGAs are procured from the same source, then possibility remains that all the resources may be implanted with HTHs having similar structure or duplicate HTHs. Thus, all three operations may produce the same erroneous result affecting system integrity, as majority polling will not be able to find the correct result in such a scenario. Hence, all the three FPGAs and related RIPs chosen for execution must be procured from different sources.

Note 1: As the structure of implanted HTHs in resources procured from diverse sources will be different, hence, probability of simultaneous trigger and same malfunction is negligible. This is established and presented in [RSK16]. Hence, only one out of three redundant operations may be affected, while the other two will work finely. The malfunctioning resource needs to be identified and replaced.

Note 2: Increasing slack time is also an objective for the present work, as additional slack time can be utilized in performing non-periodic tasks and for fault diagnosis. Slack time can be increased by executing two redundant task instances and checking their results. If they match, then no need for third redundant task execution, whose space can be considered as slack. However, if they mismatch, third redundant execution must be performed, followed by majority polling to choose the correct result.

Preventing Availability Attacks

Either simultaneously executing multiple tasks and delivering the result which arrives first before deadline or deploying checkpoints in programs and scaling up frequency of operation during anomaly is generally followed to prevent availability attacks or deadline miss [GSC17, BTW13].

Moreover, deploying checkpoints in RIPs for detecting completion is feasible. However, few issues may arise.

The present work considers infinite delays, which if takes place, the operation will be stuck at a point due to a delay loop and any amount of frequency increment in the operating platform will not be able to serve the purpose. Hence, it is necessary to detect a faulty operation and re-perform it on completely different resources for mitigating the issue related to absence of diversity, as discussed before.

Moreover, RIPs may be procured from third party vendors. In such a scenario, designing custom RIPs with checkpoints from third parties will destroy the element of anonymity and facilitate third party developers to understand the security mechanism and develop HTHs that can bypass the security techniques.

Thus, space in the schedule must be present for execution of two operations. Either one will complete within deadline and that is to be generated as output.

Re-defining Task Parameters

Preventing Integrity Attacks

As discussed previously, three task instances must be generated to protect from integrity attacks. In addition to this, a result evaluation time, $t(RE)$ and a majority polling time, $t(MP)$, must also be associated. For a task, we consider that δ_i, α_i are its deadline and arrival time respectively and ρ_i, ϵ_i are its worst case reconfiguration and execution time respectively, among its available bitstreams.

Two cases may arise:

(a) Case 1: If summation of result evaluation time, majority polling time and twice the combination of worst case execution time and maximum reconfiguration time among related RIPs for the task is less than its relative deadline, i.e. $(t(RE) + t(MP) + 2 * (\rho_i + \epsilon_i)) \leq (\delta_i - \alpha_i)$, then no frequency increment (k) in task operations are needed, or $k = 1$.

(b) Case 2: If the previous case is not valid, i.e. $(t(RE) + t(MP) + 2 * (\rho_i + \epsilon_i)) > (\delta_i - \alpha_i)$, then frequency increment has to be performed. Amount of frequency increment (k) is determined by the following formula:

$$k = \frac{(\delta_i - \alpha_i)}{(t(RE) + t(MP) + 2 * (\rho_i + \epsilon_i))} \qquad (6.1)$$

For both the cases, it must be kept in mind that though the first two task instances can execute simultaneously, execution of the third task instance is to start after completion of the first two task instances. If result mismatch occurs in the first two task instances, then only the third task instance is executed, else the space allocated for the third task instance can be considered as slack. In this space, non-periodic tasks may be scheduled or fault diagnosis operations may take place, as will be discussed later.

Preventing Availability Attacks

Space must be allocated for execution of two task instances. Like the previous one, two cases also arise here:

(a) Case 1: If summation of result checking time and twice the combination of worst case reconfiguration time and execution time of related RIPs is less than the relative deadline, i.e. $(2 * (\rho_i + \epsilon_i)) \leq (\delta_i - \alpha_i)$, then no frequency increment (k) is required, or $k = 1$, for execution of the task instances.

(b) Case 2: If Case 1 does not hold, i.e. $(2 * (\rho_i + \epsilon_i)) > (\delta_i - \alpha_i)$, then frequency must be incremented to ensure successful task completion. Amount of frequency increment (k) can be calculated as follows:

$$k = \frac{(\delta_i - \alpha_i)}{2 * (\rho_i + \epsilon_i)} \qquad (6.2)$$

Result checking is not needed here, as whichever task completes, its result is generated as output.

Like the previous scenario, it would be wise if the second task instance is executed only if the first one is unsuccessful. Thus, the second one is to start execution only after the first one completes. As for normal scenarios, execution of the second task instance will not be needed and that space can be considered as slack, which can be allocated for some non-periodic task operation or fault diagnosis.

Algorithm 1 describes task handling strategy as per security need.

Generation of Periodic Task Schedule

For periodic tasks, a resource optimized periodic task schedule needs to be generated offline, which is strictly followed at runtime.

The PTS provides details in which FPGA VP, which RIP is to be configured and at what time of the schedule, for execution of periodic task instances.

The basic mechanism for scheduling of periodic tasks is as described in Part 2 of this book. However, Part 2 Chap. 3 does not consider threats, which is our present consideration. For threat mitigation, multiple FPGAs and multiple task instances are involved, as discussed previously. As multiple instances of a particular task cannot be scheduled on the same FPGA, hence, we discuss the mechanism of allotting the different instances of a particular task in the various FPGAs in this subsection.

Note: For a single FPGA based system, this mechanism is not applicable. Only threats related to RIPs can be secured and the methodology is reduced to that of redundancy based approach, as will be discussed later in demonstration.

The mechanism is depicted via Algorithm 2, which is also described as follows:

Each FPGA can be partitioned into a fixed number of VPs, which is indexed by variable q in Algorithm 2. We start periodic task scheduling for an FPGA with a single VP or $q = 1$, from time $t = 0$ to end of schedule period.

To schedule a particular task instance in $VP(q)$, we first need to reconfigure it with related RIP via the Internal Configuration Access Port (ICAP). However, this is limited and some FPGAs posses a single ICAP and hence, checking its availability is important. If ICAP port is available, then unscheduled, available periodic task instances at time instant t is ordered as per the scheduling strategy discussed in Part 2.

Let the order for nt number of unscheduled and available task instances at time t be represented by variable p. The task instance at $p = 1$ is fetched and analyzed whether it can be scheduled for execution in $VP(q)$ of the FPGA. However, if its redundant task instance is already scheduled in any of the FPGA VPs for execution, then it is skipped to mitigate the issue of absence of diversity and the next task instance in order is analyzed. If summation of current time instant, t, worst case reconfiguration time ρ_{ij} and execution time ϵ_{ij}/k of related RIP, is less than the deadline, i.e. δ_{ij}, i.e. it can complete its execution before its deadline, then it is scheduled for execution in $VP(q)$. Else next task instance in order p is fetched and analyzed.

Algorithm 1: Task Handling as per Security Need

Input: Incoming Task T_i
T_{ij} represents jth instance of a task T_i
Output: Result of T_i, i.e. $R(T_i)$
begin

  if T_i *needs security from Integrity Attacks* **then**

    if *Case 1 holds, i.e.* $(t(RC) + 2 * (\rho_i + \epsilon_i)) \leq (\delta_i - \alpha_i)$ **then**

      Schedule T_{i1} and T_{i2}; **if** $R(T_{i1}) == R(T_{i2})$ **then**

        Generate $R(T_{i1})$ or $R(T_{i1})$ as output and consider space for T_{i3}
        as slack;

      else

        Schedule T_{i3}; **if** $R(T_{i1}) == R(T_{i3})$ **then**

          Generate $R(T_{i1})$ or $R(T_{i3})$ as output and set $Trigger(FD_k)$
          for $FPGA_k$, where T_{i2} executed;

        else

          Generate $R(T_{i2})$ or $R(T_{i3})$ as output and set $Trigger(FD_k)$
          for $FPGA_k$, where T_{i1} executed;

    else

      Case 2 holds, i.e. $(t(RE) + t(MP) + 2 * (\rho_i + \epsilon_i)) > (\delta_i - \alpha_i)$;
      Schedule T_{i1}, T_{i2} and T_{i3} and perform Majority Polling on
      $R(T_{i1}); R(T_{i2}); R(T_{i3})$;
      if $R(T_{i1}) == R(T_{i2}) == R(T_{i3})$; **then**
        Generate any one as output;

      else

        Generate the matching result as output and set $Trigger(FD_k)$
        for $FPGA_k$ which executed the task instance with anomalous
        result;

  if T_i *needs security from Availability Attacks* **then**

    if *Case 1 holds, i.e.* $(t(RC) + 2 * (\rho_i + \epsilon_i)) \leq (\delta_i - \alpha_i)$ **then**
      Schedule T_{i1};
      if $R(T_{i1})$ *available before deadline* **then**
        Generate $R(T_{i1})$ as output and consider space for T_{i2} as slack

      else

        Schedule T_{i2};
        Generate $R(T_{i2})$ as output and set $Trigger(FD_k)$ for $FPGA_k$,
        which executed T_{i1};

    else

      Case 2 holds, i.e. $(t(RC) + 2 * (\rho_i + \epsilon_i)) > (\delta_i - \alpha_i)$;
      Schedule T_{i1} and T_{i2}; **if** *Both results available before deadline* **then**
        Generate any one as output;

      else

        Generate the available result as output and set $Trigger(FD_k)$
        for the other;

On successful scheduling of a task instance in $VP(q)$, the task instance is marked scheduled and value of t is updated to time of completion of the task instance. This process is repeated from checking of the ICAP port till all task instances in the expanded task set is scheduled or till the end of the schedule period.

If schedule period ends and all task instances are not scheduled, then the process is repeated for the next FPGA VP. If scheduling has been performed for all FPGA VPs, then the process is to be repeated for another FPGA.

Algorithm 2: Offline Periodic Task Schedule Generation

Input: Expanded Task Set of Periodic Tasks
Output: Periodic Task Schedule
begin
    Set $q = 1$, where the number of FPGA VPs are denoted by variable q
    * For FPGA $VP(q)$, set $t = 0$, where the current time instant is designated
    by variable t
    for *Time Instant t* **do**
        if ** *ICAP is idle* **then**
            Available and Unscheduled task instances at time t are ordered as
            per scheduling strategy discussed in Part 2
            for *** $(p = 1; p \leq nt; p + +)$, *where p represents the order and nt*
            depicts total number of ordered, unscheduled and available task
            instances at t **do**
                Task at order p is fetched, if none of its redundant task
                instances are scheduled;
                if $(t + \rho_{ij} + \epsilon_{ij}/k) \leq \delta_{ij}$ **then**
                    Schedule the task instance for execution in $VP(q)$;
                    The task instance is marked as scheduled;
                    t is updated, i.e. $t = (t + \rho_{ij} + \epsilon_{ij}/k)$;
                    Repeat from **;
                else
                    Repeat from ***;
        else
            $t = t + 1$ (i.e. ICAP port is unavailable, hence, increment t and
            repeat from **);
    if *All task instances are not scheduled* **then**
        if *(only redundant task instances left) or (VPs not available in present*
        FPGA) **then**
            Restart execution of remaining task instances on a new FPGA from
            beginning;
        else
            Increment q, i.e. $q = q + 1$ and Restart from *;
    else
        End;

6.5.2 Online Phase

Handling Periodic Tasks

Runtime handling of periodic tasks by the agents is depicted in Algorithm 3, which is also described as follows.

On the arrival of a periodic task to $FPGA_f$ directly from an user, SAA_f analyzes its security need and based on that, it replicates the task resources. If the task needs security from integrity attacks, then three copies are made and if the task needs protection from availability attacks, two copies are made.

Based on the PTS generated offline, SAA_f dispatches the task inputs for execution. If a related task instance is scheduled to operate in its host FPGA, it sends it to the FPGA controller of its host, else outsources them to the designated FPGA as per PTS. This is done by communication with the related agent, via the SM, where the task resources are written and read by the sending and receiving agents respectively. On completion of operation, SAA_f receives the output either directly from its host FPGA controller or via a SM from another agent, which after performing related checks are delivered to the user.

The mechanism of task handling as per the security need is depicted in Algorithm 1. Only in case of an anomaly, the final task credentials is sent for execution at appropriate time. If no anomaly is associated or the scenario is normal, then no task details are sent. The receiving agent understands this and recognizes the related time period as slack and utilizes it for execution of non-periodic tasks or for fault diagnosis.

On the arrival of a periodic task instance from another agent via a SM, SAA_f passes it directly to its host FPGA controller for related placement and execution in a VP at appropriate time, as mentioned in PTS. On completion of operation, SAA_f writes back the result in the SM for collection by the sending agent.

Algorithm 3: Handling of Periodic Tasks

Input: Incoming Periodic Task: PT_i
Output: Periodic Task Output: $R(PT_i)$
begin

    for *Incoming Periodic Task from User, i.e.* PT_i **do**
        Generate task instances of PT_i, i.e. PT_{ij}, where $j = \{3, 2, 1\}$ for
        mitigating integrity attacks and $j = \{2, 1\}$ for mitigating availability
        attacks;

    Task handling as per security need is performed as per Algorithm 1, where
    the task instances are scheduled as per the PTS;

    for *Incoming Periodic Task Instances (PT_{ij}) from Other Agents* **do**
        Read Task Inputs of PT_{ij} from Shared Memory (SM);
        Schedule PT_{ij} in host FPGA at suitable time as per PTS;
        Write $R(PT_{ij})$ in SM for collection by the Sending Agent;

6.5.3 Handling Non-periodic Tasks

For non-periodic tasks, an examination time is to be associated, where it is analyzed whether a task is to be accepted or rejected. It is also determined in this time frame that if accepted, then in which FPGA VP, it is to be executed and when. Algorithm 4 depicts mechanism of non-periodic task handling, which is also described as follows.

Algorithm 4: Handling of Non-Periodic Tasks

Input: Incoming Non-Periodic Task
Output: Generate Non-Periodic Task Output
begin

    for *Incoming Non-Periodic Task from User, i.e. NPT_i* **do**
        Generate task instances of NPT_i, i.e. NPT_{ij}, where $j = \{3, 2, 1\}$ for securing from integrity attacks and $j = \{2, 1\}$ for securing from availability attacks;

        for *each task instance NPT_{ij}* **do**
            For finding available time period (ATP_{ij}) of NPT_{ij}, broadcast $\{\alpha_{ij}, \delta_{ij}\}$ and also initiate ATP_{ij} finding in itself (as per Algorithm 5);
            if *Acknowledgement received* **then**
                exit loop
            else
                repeat loop

        if *Acknowledgment not received for all NPT_i task instances* **then**
            Reject NPT_i

        else
            Accept NPT_i;
            Task handling as per security need is performed based on Algorithm 1, where each task instance NPT_{ij} is scheduled in ATP_{ij}, as received in acknowledgment;
            Note: If more than one acknowledgments are received, then NPT_{ij} is to be scheduled to the nearest FPGA;

    for *Incoming Non-Periodic Task (NPT_{ij}) from Other Agents* **do**
        Read Task Inputs of NPT_{ij} from Shared Memory;
        Schedule NPT_{ij} in ATP_{ij} as sent during acknowledgment;
        Write $Result(NPT_{ij})$ in shared memory for collection by the sending agent;

    for *Finding ATP_{ij} of NPT_{ij} from other Agents* **do**
        Read $\alpha_{ij}, \delta_{ij}, ETP_z$ from shared memory;
        Execute Algorithm 5;
        if $ATP_{ij}! = 0$ **then**
            Send ATP_{ij} as acknowledgment;

On the arrival of a non-periodic task, the agent like in the previous case, analyzes its security need and expands its task set, i.e. three task instances are created for mitigating integrity attacks and two task instances are created for mitigating availability attacks.

As no prior task schedules are present, hence, the agent broadcasts the task instances to all other agents, via the SMs. An agent on finding a non-periodic task instance, tries to find a suitable time period where it can fit the task instance in its associated FPGA. This is performed via Algorithm 5.

At end of examination time, if positive acknowledgments is not received for all the task instances, then the non-periodic task is rejected, else accepted. If accepted, then the task instances are outsourced to resources for execution via the SMs. Moreover, more than one task instance is never sent to a particular resource for execution, to mitigate the issue of absence of diversity.

Note: As no dedicated resources are considered in this work for execution of the non-periodic tasks and these needs to be accommodated in the slack spaces of the periodic task schedules, hence, a greedy approach is followed in accommodating the non-periodic task instances in the slack spaces of the periodic task schedule. Thus, three task instances for mitigating integrity attacks and two task instances for mitigating availability attacks may be simultaneously scheduled and executed, as shown in demonstration later. Thus, redefining parameters is not followed for non-periodic task instances.

Mechanism of Searching for Available Time Frame (ATP):
The methodology is illustrated in Algorithm 5 and also described below.

Searching is performed in each FPGA VP, indexed by variable q, from the arrival time of the task instance α_{ij} to its deadline δ_{ij}. t represents the time instant, *Start* and *Stop* denotes two specific time instants, while *Begin* and *End* indicates initial and terminal time of ATP_{ij} for non-periodic task NPT_{ij}.

On satisfaction of the condition at time t, *Start* is set at t, else t is incremented by 1 and rechecked, till condition is satisfied. From *Start*, a suitable time frame needs to be found where the task instance, NPT_{ij} can execute and hence, checking starts from *Start* to δ_{ij}. Thus, *Stop* is set either to δ_{ij} or till the moment when the condition becomes false. Time from *Start* to *Stop* indicates a suitable space. However, checking must be performed if ICAP port is available and the summation of reconfiguration and execution time of the related RIP for NPT_{ij} can be completed before its deadline. If these constraints meet, *Begin* and *End* are set to time instants, when NPT_{ij} can initiate and terminate its execution respectively. However, if the task instance cannot complete its operation within the available space, then the value of t is set to *Stop* and the process is repeated till the end of schedule period.

If no available period is found, then the process is to be repeated for the next VP of the FPGA. However, if no suitable space is found in all the VPs of the FPGA, i.e. $(Begin - End) = 0$, then no acknowledgment is sent.

Algorithm 5: Finding Available Time Period (ATP_{ij})

Inputs: PTS of host FPGA, Arrival Time (α_{ij}) and Deadline (δ_{ij}) of NPT_{ij}
Output: Available Time Period ATP_{ij} : $\{Begin, End, VP(q)\}$
Finding ATP_{ij} Set $q = 1$, where q indicates the number of FPGA VPs
1. **for** *FPGA VP(q)* **do**
    Set $t = \alpha_{ij}$, $Start = 0$, $Stop = 0$;
    2. **if** *(Slack present at t)* **then**
        ⌊ Set $Start = t$;
    else
        ⌊ Increment t, i.e. $t = t + 1$ and Redo Step 2;
    3. **for** $t = Start$ to δ_{ij} **do**
        if $t == \delta_{ij}$ **then**
        ⌊ Set $Stop = \delta_{ij}$

        else if *(Slack present at t)* **then**
        ⌊ Increment t, i.e. $t = t + 1$ and Repeat Step 3;

        else
        ⌊ Set $Stop = t$
    4. **for** $t = Start$ to $Stop$ **do**
        if $((t + \rho_{ij} + \epsilon_{ij}) \le \delta_{ij})$ && *(ICAP port available from t to $(t + \rho_{ij})$)* **then**
        Set $Begin = t$,
        $End = (t + \rho_{ij} + \epsilon_{ij})$,
        Generate output ATP_{ij} : $\{Begin, End, VP(q)\}$;
        ⌊ Exit loop and end;

        else
        ⌊ Increment t, i.e. $t = t + 1$ and Repeat from Step 4;
    5. **if** $Stop < \delta_{ij}$ **then**
        ⌊ Set $t = Stop$ and Repeat from Step 2;
6. **if** *more FPGA VPs available* **then**
  ⌊ Set $q = q + 1$ and restart from Step 1

6.5.4   Fault Handling

Algorithm 6 depicts the mechanism, which is also discussed below.

Either the RIP or the FPGA resource can be affected. For this, when $Trigger(FD_k)$ is set, SAA_k performs the same operation in the same VP of its host FPGA during free time, but with a RIP procured from a different source. If same anomalous result is obtained, then the FPGA is affected, but if a different result is obtained, then the RIP is affected.

If the FPGA is affected, SAA_k generates an alarm for its replacement. In the time period while the FPGA is not replaced, all the existing scheduled tasks whose redundant task instances are not available, specially for the non-periodic tasks, the agent outsources them to other agents for execution. Moreover, if some free FPGA resources are present in the system, then the agents directly send the task instances to the agent of the free resources for their execution. However, if redundant task instances are available like the periodic tasks, then SAA_k informs the other agents, which in turn executes the redundant task instances that were initially considered as slacks.

If the RIP is found affected, then it is necessary to prevent the use of resources procured from the malicious vendor, to mitigate the issue of duplicated HTHs. Hence, the source or RIP vendor is blacklisted and this information is broadcasted to all other agents via the SMs.

Algorithm 6: Fault Handling by SAA

Input: Trigger(FD_k)
Output: $FPGA_k$ affected or RIP affected
if $Trigger(FD_k)$ *is set* **then**

> SAA_k schedules same operation in same FPGA VP in free time but with RIP procured from different vendor;
>
> **if** *Same anomalous result obtained* **then**
>
>> $FPGA_k$ is affected:
>> SAA_k raises alarm for change of $FPGA_k$
>> If redundant task instances are not available, then SAA_k outsources them to other agents for execution, else informs other agents to execute the redundant task instances that were initially considered as slacks;
>> This is performed till $FPGA_k$ is replaced;

else

> RIP is affected:
> SAA_k blacklists vendor who supplied RIP and sends this information to other agents via SMs;

6.5.5   Demonstration

For demonstration, we consider the real time task schedules depicted in Fig. 6.4.

Note: There is a difference between original slacks and potential slacks. Unutilized spaces that are present in the original schedule are termed as slacks or original slacks, while potential slacks are spaces that can be considered as free or slacks if their redundant task instances executed successfully. Thus, the potential slacks can only be considered for scheduling after its redundant task instances executed successfully. This is depicted in the pictures of demonstrations.

Fully Reconfigurable Single FPGA System

The periodic schedules for mitigating integrity and availability attacks are similar to that described for redundancy based approach and as depicted diagrammatically in Fig. 6.10.

For normal scenario, the space allocated for third sub instance for mitigating integrity attacks, i.e. T_{ijc} and the second task instance for mitigating availability attacks, i.e. T_{ijb} remains normally free and hence, can be considered as slack, as depicted in Fig. 6.14a and b respectively, for fully re-configurable systems.

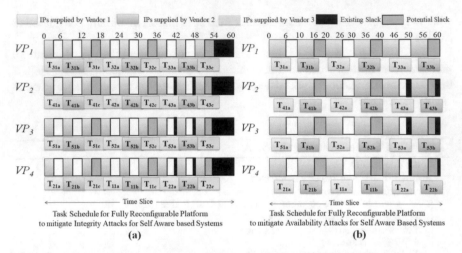

Fig. 6.14 Depicting slacks for normal scenario of task schedules for fully re-configurable platforms via self aware mechanism for **a** mitigating integrity attacks **b** mitigating availability attacks

Let two non-periodic tasks, NPT_1 and NPT_2 arrive at time instant t = 10 and deadline at t = 30, with execution time of 14 time units. On arrival, the self aware agents checks their security need. Three subtask instances are created if it needs security from integrity attack ($NPT_{1a}, NPT_{1b}, NPT_{1c}$ for task NPT_1 and $NPT_{2a}, NPT_{2b}, NPT_{2c}$ for task NPT_2), and two subtask instances are created if it needs security from availability attack (NPT_{1a}, NPT_{1b} for task NPT_1 and NPT_{2a}, NPT_{2b} for task NPT_2). Suitable ATP is searched in the FPGA VPs for allocation of the non-periodic task instances. Based on the available time slacks, appropriate frequency increment is performed so that the task instances can complete their execution in the respective slack spaces.

For full re-configurable systems, considering normal scenario and no associated threat is present, slack spaces of T_{31c}, T_{41c} and T_{51c} is used to execute NPT_{1a}, NPT_{1b}, NPT_{1c} for mitigating integrity attacks. Thus, NPT_1 is accepted and scheduled. However, for NPT_2, only slack space for NPT_{2a} is found, but not for NPT_{2b} and NPT_{2c}. Hence, it is rejected. This is shown in Fig. 6.15a. However, as two task instances are sufficient for mitigating availability attacks, hence, NPT_{1a}, NPT_{1b} of NPT_1 are scheduled in slack spaces of T_{31b} and T_{41b} respectively, while NPT_{2a}, NPT_{2b} of NPT_2 are scheduled in slack spaces of T_{31b}, T_{21b} respectively. This is as shown in Fig. 6.15b.

Now, let's consider an attack scenario. Let bitstream associated with T_{42b} is associated with a malfunction, as depicted in Fig. 6.15a. As, results mismatch for T_{42a} and T_{42b}, hence, T_{42c} is executed. Via majority polling, it is found that T_{42b} is associated with a malfunction. As, this is a single FPGA based system, hence, vulnerability is associated with the bitstream. Thus, from next phase, bitstreams or IPs supplied by vendor 2 is ignored and replaced with bitstreams or IPs supplied by vendor 4. This is illustrated in Fig. 6.15a.

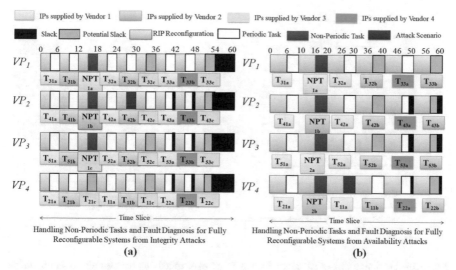

Fig. 6.15 Handling non-periodic tasks and fault diagnosis for fully reconfigurable systems from **a** integrity attacks **b** availability attacks

Similarly, let's consider that T_{11a} cannot complete execution in time, as depicted in Fig. 6.15b. Hence, T_{11b} is executed and the space for T_{11b} cannot be considered as slack. As, T_{11b} completes execution within time, then it is concluded that vulnerability is associated with bitstream or IP procured from T_{11a}. Thus, from the next phase onwards, IPs procured from vendor 1 is ignored and replaced with IPs procured from vendor 4 (vendor 3 is also possible), as illustrated in Fig. 6.15b.

Partially Reconfigurable Single FPGA System

Figure 6.16 depicts the periodic schedules for mitigating integrity and availability attacks. For demonstration, we take the aid of this example. For normal scenario, i.e. when no attack takes place, space assigned for execution of τ_{ijc} for mitigating integrity attacks, as shown in Fig. 6.16a can be considered as slacks. Similarly, space assigned for execution of τ_{ijb} for mitigating availability attacks can be considered as slacks, as depicted in Fig. 6.16b These spaces can be used for scheduling of non-periodic tasks or for fault diagnosis.

Like fully single FPGA re-configurable platforms, let us also consider in this case that two non-periodic tasks, NPT_1 and NPT_2. However, if deadline of the non-periodic tasks is at t = 30, then neither will be available as no suitable slack is present. However, if their deadline is at t = 40, then execution is possible. For the case of mitigating integrity attacks, three task instances need to be generated. NPT_{1a}, NPT_{1b}, NPT_{1c} will be scheduled in place of τ_{11c}, τ_{22c}, τ_{41c} respectively. Thus, NPT_1 will be accepted, but NPT_2 won't be accepted, due to lack of slack spaces. This is depicted in Fig. 6.17a. For the case of mitigating availability attacks, two task instances need to be generated. NPT_{1a}, NPT_{1b} will be scheduled in place of τ_{11b}, τ_{22a} respectively, while NPT_{2a}, NPT_{2b} will be scheduled in place of τ_{41b}, τ_{52b} respectively. Both NPT_1 and

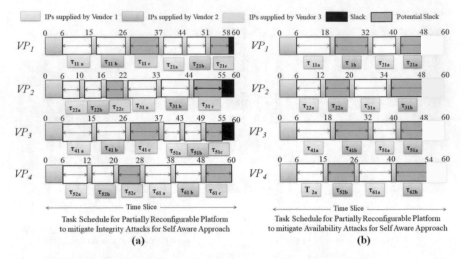

Fig. 6.16 Depicting slacks for normal scenario of task schedules for partially re-configurable platforms via self aware mechanism for **a** mitigating integrity attacks **b** mitigating availability attacks

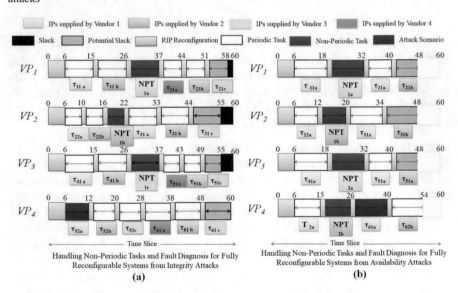

Fig. 6.17 Handling non-periodic tasks and fault diagnosis for partially reconfigurable systems from **a** integrity attacks **b** availability attacks

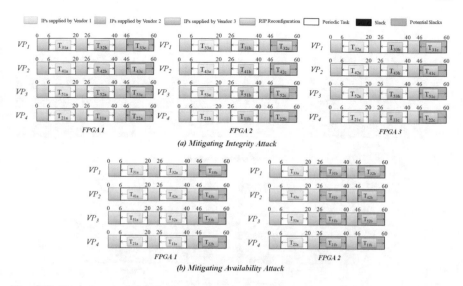

Fig. 6.18 Depicting potential slacks for normal scenario for fully re-configurable multi FPGA systems from **a** integrity attacks **b** availability attacks

NPT_2 will be accepted for execution. However, unlike the previous case, frequency increment of related task instances in the present case will be different, as length of the slack spaces is also different. This is depicted in Fig. 6.17b.

For demonstrating attack scenario, let's consider bitstream associated with τ_{52a} of Fig. 6.17a and τ_{61a} of Fig. 6.17b is associated with a malfunction. Results will mismatch in the former, while task execution cannot be completed in the latter. Thus, space allocated for τ_{52c} and τ_{62b} won't be considered as slacks and task executions as pre-decided will take place, as depicted in Fig. 6.17. When malfunction is confirmed in a bitstream, then the vendor from which it is procured is blacklisted and in future operations, the vendor is changed, as depicted in Fig. 6.17, where on confirmation that τ_{52a} executed maliciously, vendor 1 is blacklisted and its supplied bitstreams is replaced with bitstreams from vendor 4 for future operations.

Multi FPGA Fully Re-configurable Systems

For multi FPGA fully re-configurable systems, scheduling of task instances to mitigate integrity attacks and availability attacks is depicted in Fig. 6.18a and b respectively. It is to be noted that neither bitstreams procured from the same vendor, nor the same FPGA platform is used for scheduling execution of instances of a particular task. As only during an anomaly, the third task instance is executed, hence, for normal scenarios, space and time dedicated for the third task instance can be considered as a potential slack, as shown in Fig. 6.18.

Handling of non-periodic tasks and attacks is depicted in Fig. 6.19.

Let two non-periodic tasks, NPT_1 and NPT_2 arrive. Arrival time and deadline of NPT_1 is at t = 35 and t = 60 respectively, while that of NPT_2 be at t = 15 and t = 45 respectively. For mitigating integrity attacks, three task instances are generated.

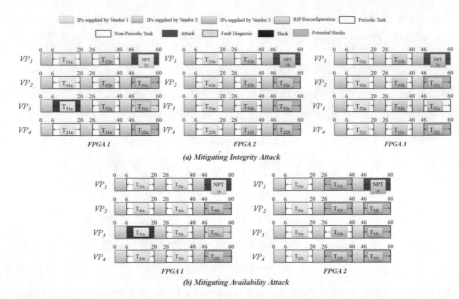

Fig. 6.19 Handling non-periodic tasks and fault diagnosis for fully re-configurable multi FPGA systems from **a** integrity attacks **b** availability attacks

Slack spaces for executing all the task instances are available for NPT_1, but not for NPT_2, hence NPT_1 is accepted, while NPT_2 is rejected, as depicted in Fig. 6.19a. For mitigating availability attacks, two task instances are generated. Though slack spaces are available for executing all task instances of NPT_1, but space for only one task instance is available for NPT_2. Hence, NPT_1 is accepted but NPT_2 is rejected, as depicted in Fig. 6.19b.

Now, let's consider an attack scenario.

Initially let's consider integrity attack. For this, let us consider that result mismatch occur for T_{51a} and T_{51b}. Hence, T_{51c} has to be executed. On performing majority polling, it is found that T_{51b} and T_{51c} generate identical results. Hence, anomaly is associated with operation of T_{51a}. Hence, fault diagnosis has to be performed. For fault diagnosis, bitstream associated with operation of T_{51c} is executed in the same VP of FPGA1 in slack space. This is depicted in Fig. 6.19a.

Now, let's consider availability attacks. For this, let's assume that T_{51a} cannot complete its operation in its designated time slice. Hence, T_{51b} is executed, and fault diagnosis is performed with bitstream of T_{51c} in the slack time available. This is shown in Fig. 6.19b.

In fault diagnosis, if result mismatch still occurs, then fault is associated with bitstream and for future operations, vendor from which the faulty RIP is procured, which in this case is vendor 1, is blacklisted and all RIPs procured from it is not used in future operations. RIPs from another vendor is used, which in the present case is from vendor 4, as depicted in Fig. 6.20. Else, fault is associated with the FPGA, and operations of it is halted and alarm is raised for its replacement. Until replaced,

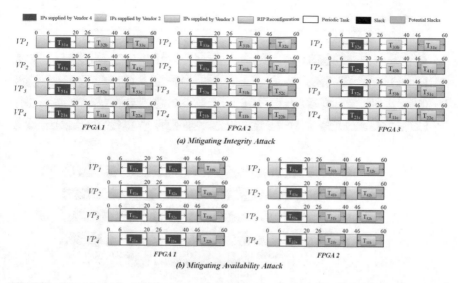

Fig. 6.20 Scenario when anomaly is associated with bitstreams or RIPs for fully re-configurable multi FPGA systems from **a** integrity attacks **b** availability attacks

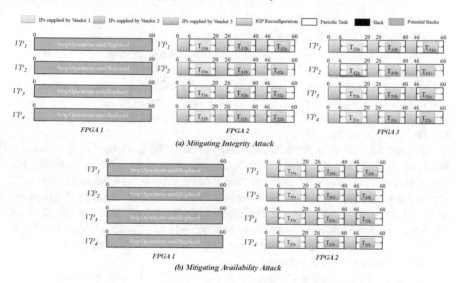

Fig. 6.21 Scenario when anomaly is associated with an FPGA for fully re-configurable multi FPGA systems from **a** integrity attacks **b** availability attacks

it's operations may be either outsourced to other FPGAs, if redundant operations are not available (not shown in current Figure). Else, the redundant operations that were initially considered as slacks are activated, as shown in Fig. 6.21.

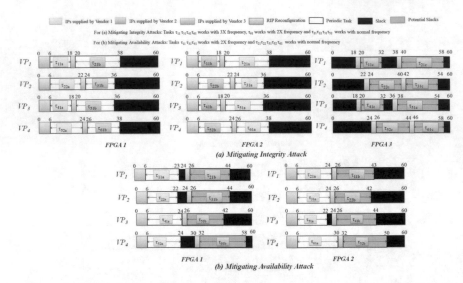

Fig. 6.22 Depicting potential slacks for normal scenario for partially re-configurable multi FPGA systems from **a** integrity attacks **b** availability attacks

Multi FPGA Partially Re-configurable Systems

Periodic task schedules for mitigating integrity and availability attacks for multi FPGA partially reconfigurable platforms is depicted in Fig. 6.22a and b respectively. Like the previous scenarios, third task instance for mitigating integrity attacks and second task instance for mitigating availability attacks can be considered as potential slacks for normal scenarios. These can be used for execution of non-periodic tasks and for fault diagnosis.

Let's consider two non-periodic tasks, NPT_1 and NPT_2. Let arrival time, execution time and deadline for NPT_1 be at t = 35, 15 time units and t = 60 respectively. For NPT_2, arrival time, execution time and deadline be at t = 0, 10 time units and t = 20 respectively. As slack spaces are available for NPT_1, but not for NPT_2 for both the cases, i.e. mitigating integrity and availability attacks, hence, NPT_1 is accepted for execution and scheduled in slack spaces as depicted in Fig. 6.23. However, NPT_2 is rejected due to unavailability of suitable slacks for all of its task instances.

Let's consider an attack scenario.

For the case of integrity attacks, let result mismatch occurs for τ_{41a} and τ_{41b}. Hence, τ_{41c} is scheduled as depicted in Fig. 6.23a. After majority polling, if results of τ_{41c} and τ_{41b}, matches, then it can be confirmed that anomaly is associated with operation of τ_{41a}. Thus, after completion of τ_{41c}, fault diagnosis is performed in FPGA1, where an alternative RIP performing the same task, which in the present case is RIP procured from vendor 3, i.e. τ_{41c} is executed in the same VP of FPGA1, as depicted in Fig. 6.23a.

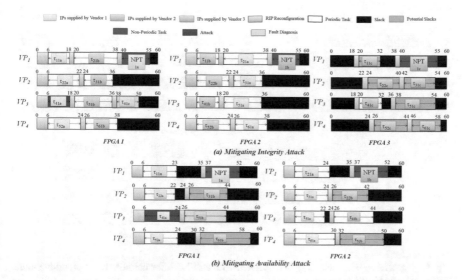

Fig. 6.23 Handling non-periodic tasks and fault diagnosis for partially re-configurable multi FPGA systems from **a** integrity attacks **b** availability attacks

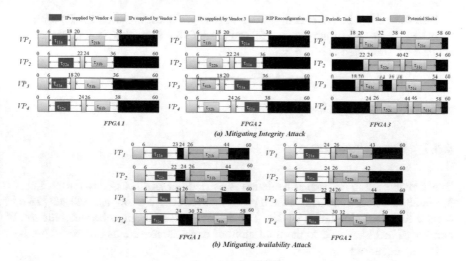

Fig. 6.24 Scenario when anomaly is associated with bitstreams or RIPs for partially re-configurable multi FPGA systems from **a** integrity attacks **b** availability attacks

For the case of availability attacks, if τ_{41a} cannot complete before its designated time, then τ_{41b} is scheduled for execution. For fault diagnosis, τ_{41b} is re-executed in VP3 of FPGA1. This is shown in Fig. 6.23b.

If results mismatch, then the RIP associated with τ_{41a} is associated with anomaly. Thus, the vendor from which it is procured, i.e. vendor 1 is blacklisted and alternative RIPs from another vendor, i.e. vendor 4 is used in future operations as depicted in Fig. 6.24.

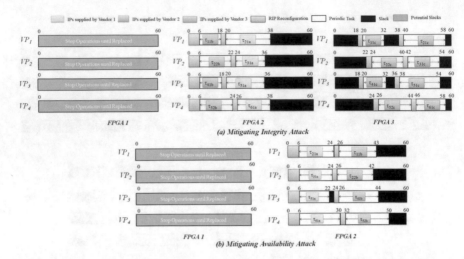

Fig. 6.25 Scenario when anomaly is associated with an FPGA for partially re-configurable multi FPGA systems from **a** integrity attacks **b** availability attacks

Else the FPGA is associated with an anomaly. Alarm is raised for its replacement and its operations are stopped till replaced. Other redundant task instances are activated, as depicted in Fig. 6.25.

6.6 Experimentation and Results

6.6.1 Experimentation

Experimental setup and experimentation is carried out as discussed in Part 2, Chap. 3. Simulation based experiments are performed. Metric *Task Rejection Rate (TRR)* is used for evaluation and analysis. *TRR* is the percentage ratio between the total number of tasks rejected to the total number of tasks arrived, as expressed by the following equation:

$$TRR = (\nu/\psi) \times 100 \qquad (6.3)$$

where, ν and ψ denote the total number of tasks rejected and total number of tasks arrived, respectively.

However, threat analysis was not performed before. Threat analysis for the present case is associated due to two factors. First, vulnerability related to bitstreams or RIPs, where the supplying vendor may implant malicious codes. Second, threat related to malicious FPGA fabric, where adversaries in the foundry may implant malicious elements in the FPGA fabric.

For experimentation of threat due to bitstreams or RIPs, malicious verilog codes are inserted during the hardware description language (HDL) phase of RIP design.

As the malicious codes are quite small in number as compared to the original codes, hence, their detection is difficult. For experimentation of threat due to FPGAs, a small virtual portion is created in the FPGA fabric, where malicious logic is embedded. This is same as discussed in the previous chapter. However, the structure of the payload differs. Presently, for causing integrity attacks, additional functionality is provided by appending logic gates that generates erroneous results and for causing availability attacks, several buffers with loop architecture are provided that causes finite and infinite delays.

6.6.2 Result Analysis

For fully reconfigurable systems, the normal scenario is presented in Part 2, Chap. 3. This is reproduced here in Fig. 6.26 for quick reference. As utilization increases, TRR increases. This is due to the fact that as the number of processing elements or VPs is fixed and can process only a limited number of tasks, hence, as the number of tasks increases, TRR also increases. When μ (average individual task weight) increases, TRR increases. As for a fixed number of VPs and for fixed utilization, increase in μ, reduces the number of tasks that can be executed and hence, TRR reduces.

When attacks occur, which may be either integrity or availability, either erroneous results are generated or no results are generated within time. Hence, task success does not take place. This can alternatively be represented as high task rejection. Thus, when utilization increases, TRR increases more than the normal scenario. However, the rate of TRR varies with the change in utilization. This is due to the fact that as the number of VPs are fixed, hence, when utilization is less, system slacks is more and hence, effect of threat is less. But when utilization is more, system slacks is less and hence, effect of threat increases that adversely increases the TRR of the system. This is demonstrated in Fig. 6.27.

When self aware mitigation strategy as discussed is applied, conditions become better but does not reach the ideal case, which is the normal scenario for the present work. This is due to the fact that with detection of anomaly and fault diagnosis, faulty resources that can be either the FPGAs or bitstreams are replaced. If anomaly occurs early, i.e. when utilization is low, then related detection and replacement of faulty resources is also made early, and the scenario becomes almost equal to the normal scenario. But if anomaly occurs late, i.e. when utilization is high, then in many cases detection and replacement of faulty resources is not possible, which adversely increases TRR. This is depicted in Fig. 6.28.

For partially reconfigurable systems, the trend is similar to that of fully reconfigurable systems, i.e. with increase in utilization, TRR increases and with increase in μ, TRR decreases. However, as partial reconfiguration overhead is less and as greater flexibility in scheduling is possible than fully reconfigurable systems, hence, TRR is less in partially reconfigurable systems than TRR for fully reconfigurable systems. Moreover, for partially reconfigurable systems, reconfiguration of all the VPs can be made asynchronously, hence, TRR is low than fully reconfigurable systems. Figure 6.29 depicts the scenario via a bar graph.

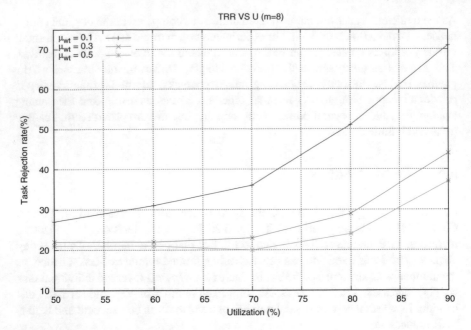

Fig. 6.26 Normal scenario for fully reconfigurable systems

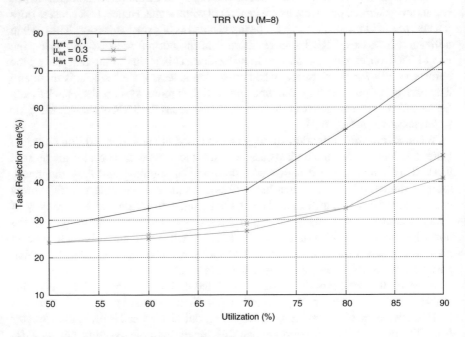

Fig. 6.27 Threat scenario for fully reconfigurable systems

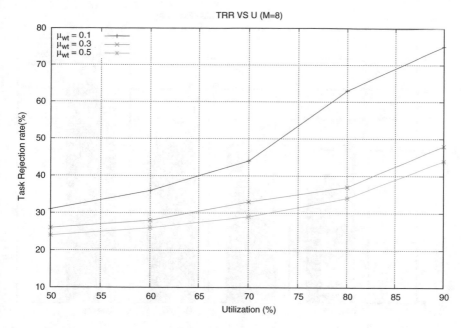

Fig. 6.28 Mitigation for fully reconfigurable systems

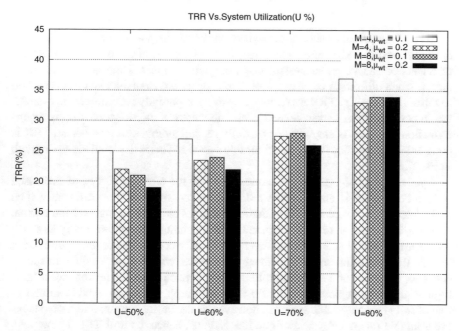

Fig. 6.29 Normal scenario for partially reconfigurable systems

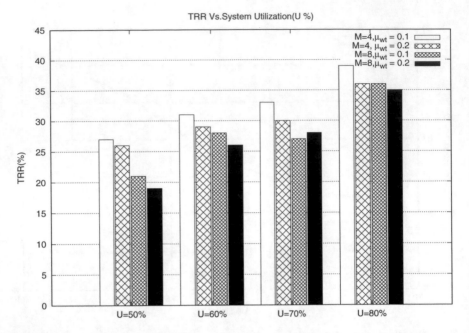

Fig. 6.30 Threat Scenario for partially reconfigurable systems

When threat occurs, scenarios is almost similar to the previous case, i.e. for fixed M and μ and low utilization scenarios, more slacks are available. Hence, effects of threats or attacks on successful task completion is less. However, for fixed M and μ, when utilization increases, less slack spaces are available, so effects related to threats is more, i.e. TRR increases. However, as partially reconfigurable systems are in consideration, hence, related available slacks is more. Moreover, as reconfiguration overhead is also low for partially reconfigurable systems, hence, TRR is comparatively low than fully reconfigurable systems. This is depicted via a bar graph in Fig. 6.30.

Application of mitigation technique decreases the effect of threat, as discussed before. For a fixed M and μ, when utilization is low or anomaly occurs early, then early detection and replacement of faulty resources aid in reaching near the normal scenario, i.e. deviation of TRR when mitigation technique is applied is very less than TRR for the normal scenario. However, if anomaly occurs late or when utilization is high, then detection and replacement of resources may not be possible. In such a scenario, deviation of TRR when mitigation is applied from the normal scenario is high. Figure 6.31 depicts the scenario via a bar graph. Thus, the trend is similar to that of fully reconfigurable systems, however, as flexibility is more and more slacks are available for partially reconfigurable systems, hence, overall TRR is low than fully reconfigurable systems.

Analysis for fully reconfigurable multi FPGA systems with respect to increase in utilization percentage and increment in number of FPGAs is depicted graphically in

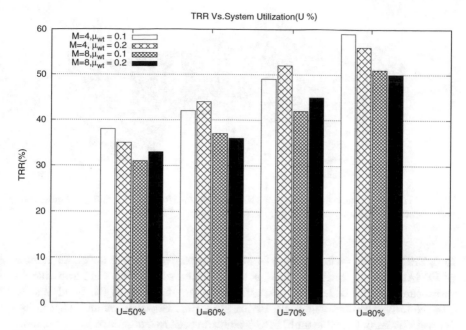

Fig. 6.31 Mitigation for partially reconfigurable systems

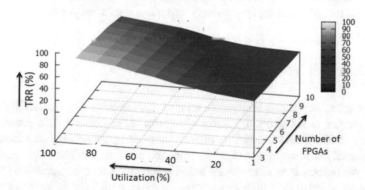

Fig. 6.32 Graphical representation of TRR for multi FPGA fully reconfiguration systems with respect to utilization percentage and increase in number of FPGAs

Fig. 6.32. As slacks reduces with increment in utilization, hence, TRR increases with increase in utilization percentage. This is discussed previously. Moreover, as evident from the graphical analysis, it is clear that with increase in number of FPGAs, TRR decreases. This is due to the fact that more resources and slacks are available for re-executing a task which has been affected by an attack.

Figure 6.33 provides a graphical analysis for partially reconfigurable multi FPGA systems with respect to increase in utilization percentage and increment in number of FPGAs. Like the previous scenario, as slacks reduces with increment in utiliza-

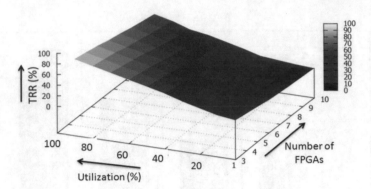

Fig. 6.33 Diagrammatic representation of TRR for multi FPGA partially reconfiguration systems with respect to utilization percentage and increase in number of FPGAs

tion, TRR enhances with increase in utilization percentage. Moreover, as number of FPGAs increase, TRR decreases, as more resources and slacks are available for re-executing a task which has been affected by an attack. Thus the trend is same like the fully reconfigurable multi FPGA platforms. However, overall TRR is low for partially reconfigurable multi FPGA platforms due to the same reasons, i.e. more flexibility is present for task scheduling in partially reconfigurable systems as the VPs can be reconfigured asynchronously and more slacks are present.

6.7 Conclusion

In this chapter, we discuss how active attacks, i.e. integrity and availability attacks may jeopardize task scheduling. Integrity attacks are associated with generation of erroneous results, while availability attacks prevents result generation within deadline. We analyze how vulnerability in both the RIPs or FPGAs may cause such attacks. We discuss how redundancy based mechanism may be applied to the current context and also state their limitations. To remove such limitations, we propose self aware strategies, where we develop low overhead self aware agents that work based on the observe-decide-act principle. We depict handling of both periodic and non-periodic tasks and also fault diagnosis on detection of threat. Simulation based experiments are carried out for experimentation. Graphical analysis is performed for results, based on the metric task rejection rate with respect to utilization percentage and increase in number of FPGAs for a particular system.

References

[AAS14] H.A.M. Amin, Y. Alkabani, G.M.I. Selim, System-level protection and hardware Trojan detection using weighted voting. J. Adv. Res. **5**(4), 499–505 (2014)

[BHBN14] S. Bhunia, M.S. Hsiao, M. Banga, S. Narasimhan, Hardware trojan attacks: threat analysis and countermeasures. Proc. IEEE **102**(8), 1229–1247 (2014)

[BM13] C. Bolchini, A. Miele, Reliability-driven system-level synthesis for mixed-critical embedded systems. IEEE Trans. Comput. **62**(12), 2489–2502 (2013)

[BT18] S. Bhunia, M. Tehranipoor, *Hardware Security—A Hands on Approach* (Elsevier Morgan Kaufmann Publishers, 2018). ISBN: 9780128124772

[BTW13] M.-S. Bouguerra, D. Trystram, F. Wagner, Complexity analysis of checkpoint scheduling with variable costs. IEEE Trans. Comput. **62**(6), 1269–1275 (2013)

[CD07] K. Chatterjee, D. Das, Semiconductor manufacturers' efforts to improve trust in the electronic part supply chain. IEEE Trans. Compon. Packag. Technol. **30**(3), 547–549 (2007)

[CMSW14] X. Cui, K. Ma, L. Shi, K. Wu, High-level synthesis for run-time hardware Trojan detection and recovery, in *Proceedings of the 51st Annual Design Automation Conference, DAC '14* (2014), pp. 157:1–157:6

[GDT14] U. Guin, D. DiMase, M. Tehranipoor, Counterfeit integrated circuits: detection, avoidance, and the challenges ahead. J. Electron. Test. **30**(1), 9–23 (2014)

[GMSC18] K. Guha, A. Majumder, D. Saha, A. Chakrabarti, Reliability driven mixed critical tasks processing on FPGAS against hardware Trojan attacks, in *21st Euromicro Conference on Digital System Design, DSD 2018*, Prague, Czech Republic, August 29–31, 2018, ed. by M. Novotný, N. Konofaos, A. Skavhaug (IEEE Computer Society, 2018), pp. 537–544

[GMSC19] K. Guha, A. Majumder, D. Saha, A. Chakrabarti, Criticality based reliability against hardware trojan attacks for processing of tasks on reconfigurable hardware, Microprocess. Microsyst. **71** (2019)

[GSC17] K. Guha, D. Saha, A. Chakrabarti, Self aware SoC security to counteract delay inducing hardware Trojans at runtime, in *2017 30th International Conference on VLSI Design and 2017 16th International Conference on Embedded Systems (VLSID)* (2017), pp. 417–422

[GSC18] K. Guha, S. Saha, A. Chakrabarti, Shirt (self healing intelligent real time) scheduling for secure embedded task processing, in *2018 31st International Conference on VLSI Design and 2018 17th International Conference on Embedded Systems (VLSID)* (2018), pp. 463–464

[GSC19a] K. Guha, D. Saha, A. Chakrabarti, SARP: self aware runtime protection against integrity attacks of hardware Trojans, in *VLSI Design and Test*, Singapore (2019), pp. 198–209

[GSC19b] K. Guha, D. Saha, A. Chakrabarti, Stigmergy-based security for SoC operations from runtime performance degradation of SoC components. ACM Trans. Embed. Comput. Syst. **18**(2), 14:1–14:26 (2019)

[GZFT14] U. Guin, X. Zhang, D. Forte, M. Tehranipoor, Low-cost on-chip structures for combating die and IC recycling, in *Proceedings of the 51st Annual Design Automation Conference, DAC '14* (2014), pp. 87:1–87:6

[HKM+14] T. Hayashi, A. Kojima, T. Miyazaki, N. Oda, K. Wakita, T. Furusawa, Application of FPGA to nuclear power plant I&C systems, in *Progress of Nuclear Safety for Symbiosis and Sustainability* (Springer, 2014), pp. 41–47

[KPK08] T.H. Kim, R. Persaud, C.H. Kim, Silicon odometer: an on-chip reliability monitor for measuring frequency degradation of digital circuits. IEEE J. Solid-State Circuits **43**(4), 874–880 (2008)

[LJ01] C. Liu, J. Jou, Efficient coverage analysis metric for HDL design validation. IEE Proc. Comput. Digit. Tech. **148**(1), 1–6 (2001)

[LRYK13] C. Liu, J. Rajendran, C. Yang, R. Karri, Shielding heterogeneous MPSoCs from untrustworthy 3PIPs through security-driven task scheduling, in *2013 IEEE International Symposium on Defect and Fault Tolerance in VLSI and Nanotechnology Systems (DFTS)* (2013), pp. 101–106

[LRYK14] C. Liu, J. Rajendran, C. Yang, R. Karri, Shielding heterogeneous MPSoCs from untrustworthy 3PIPs through security-driven task scheduling. IEEE Trans. Emerg. Top. Comput. **2**(4), 461–472 (2014)

[MKN+16] S. Mal-Sarkar, R. Karam, S. Narasimhan, A. Ghosh, A. Krishna, S. Bhunia, Design and validation for FPGA trust under hardware Trojan attacks. IEEE Trans. Multi-Scale Comput. Syst. **2**(3), 186–198 (2016)

[MWP+09] D. McIntyre, F. Wolff, C. Papachristou, S. Bhunia, D. Weyer, Dynamic evaluation of hardware trust, in *2009 IEEE International Workshop on Hardware-Oriented Security and Trust* (2009), pp. 108–111

[NDC+13] S. Narasimhan, D. Du, R.S. Chakraborty, S. Paul, F.G. Wolff, C.A. Papachristou, K. Roy, S. Bhunia, Hardware Trojan detection by multiple-parameter side-channel analysis. IEEE Trans. Comput. **62**(11), 2183–2195 (2013)

[RSK16] J.J. Rajendran, O. Sinanoglu, R. Karri, Building trustworthy systems using untrusted components: a high-level synthesis approach. IEEE Trans. Very Large Scale Integr. (VLSI) Syst. **24**(9), 2946–2959 (2016)

[SDG+15] S. Sarma, N. Dutt, P. Gupta, N. Venkatasubramanian, A. Nicolau, Cyberphysical-system-on-chip (CPSoC): a self-aware MPSoC paradigm with cross-layer virtual sensing and actuation, in *Proceedings of the 2015 Design, Automation, Test in Europe Conference & Exhibition, DATE '15* (2015), pp. 625–628

[TK10] M. Tehranipoor, F. Koushanfar, A survey of hardware Trojan taxonomy and detection. IEEE Des. Test Comput. **27**(1), 10–25 (2010)

[XFT14] K. Xiao, D. Forte, M. Tehranipoor, A novel built-in self-authentication technique to prevent inserting hardware Trojans. IEEE Trans. Comput. Aided Des. Integr. Circuits Syst. **33**(12), 1778–1791 (2014)

[XZT13] K. Xiao, X. Zhang, M. Tehranipoor, A clock sweeping technique for detecting hardware Trojans impacting circuits delay. IEEE Des. Test **30**(2), 26–34 (2013)

[ZWB15] Y. Zheng, X. Wang, S. Bhunia, SACCI: scan-based characterization through clock phase sweep for counterfeit chip detection. IEEE Trans. VLSI Syst. **23**(5), 831–841 (2015)

Chapter 7
Handling Power Draining Attacks

7.1 Introduction

With the advent of Industry 4.0 or the fourth industrial revolution [RB18], a key issue was to ensure flexibility and reconfigurability of the processing resources. Moreover, direct task execution in hardware enhances speed of operation and also ensures security from various types of software attacks. Previous hardware based systems, i.e. application specific integrated circuits (ASICs), though ensured high speed in task operation, but did not possess flexibility, i.e. different types of tasks could not be executed in the same platform. Such a limitation was eradicated with advent of reconfigurable hardware or field programmable gate arrays (FPGAs) [Xil18]. With its ability of dynamic partial reconfiguration at runtime, FPGAs provided the necessary flexibility for execution of different types of tasks on the same platform, in addition to hardware acceleration and security from various types of software attacks [Xil10, Xil18].

Thus, FPGAs found wide usage in embedded and cyber physical platforms, where an important factor lies in execution of varied tasks on the same platform [HKM+14]. As optimizing hardware resources is an important facet of such systems. In addition to this, cyber physical systems are associated with sensors and the action undertaken in due course is based on the feedback provided by sensors [XGS+15]. As FPGAs possess the ability to change its operations with respect to time, hence, FPGAs are a perfect fit for such arenas.

An important aspect of such embedded and cyber physical systems is that they are associated with strict energy constraints or pre-specified power budgets [KM18]. Thus, the objective for such systems is to ascertain completion of all periodic tasks and facilitate execution of a maximum non-periodic tasks within the fixed energy or power budget.

Though task operations directly on hardware eradicates various software threats, but with entry into the embedded arena, hardware threats has also been witnessed [BHBN14, TK10]. For task execution on reconfigurable hardware platforms, bitstreams or reconfigurable intellectual properties (RIPs) need to be procured from

© The Author(s), under exclusive license to Springer Nature Switzerland AG 2021
K. Guha et al., *Self Aware Security for Real Time Task Schedules in Reconfigurable Hardware Platforms*, https://doi.org/10.1007/978-3-030-79701-0_7

various third party IP (3PIP) vendors. These are related to various task executions and such RIPs essentially configures the FPGA fabric for execution of the related task. Vulnerabilities may be present in either the RIPs or the FPGA fabric. Malicious 3PIP vendors may insert malicious codes in the supplied RIPs [RSK16, LRYK14]. Implantation of malicious circuitry in unused spaces of a layout by adversaries in the foundry during FPGA fabrication [XFT14], may result in malfunctioning of task operations associated with the FPGA platform. These are collectively termed as Hardware Trojan Horses (HTHs). HTHs are particularly dangerous due to its property of remaining dormant or in a sleeping mode during testing and initial stages of operations. But with satisfaction of trigger (which may be either external or internal) at runtime, they get activated and jeopardizes runtime operations [Boa05, BHBN14]. Details of structure of HTHs are discussed in previous chapters.

HTH attacks may lead to various active and passive threats. Active threats involve either denial of service [GSC19b] or generation of incorrect or erroneous results [MWP+09, GSC19a]. The former is associated with affecting integrity of the system, while the latter jeopardizes availability of the system. Passive attacks affect user privacy by leaking secret information via covert channels [LJM13, GSC17], that leads to lack of user trust and reliability. These we have discussed in the previous two chapters and even showed how self aware strategies is useful to overcome such issues.

In the present chapter, we focus on power draining attacks of HTHs. Such attacks may cause the system to cease its functional operations before its expected lifetime, or simply reduce the lifetime of the system. Critical operations scheduled during middle or end phases may not take place in such scenarios, that may lead to fatal consequences.

As discussed previously, such systems are associated with executions of multiple tasks, arranged in schedules. As previous focus of security strategies was confined to software, researchers considered the hardware to be trusted and proposed efficient energy or power aware scheduling strategies that facilitated maximum task execution within pre-determined energy or power budget [SP18, BMAB16]. Even reliability driven task scheduling strategies from faults and software attacks were even proposed [Tos12, MFP16]. But these do not focus on hardware related threats. Thus, analysis of power draining attacks due to vulnerability of hardware on real time task schedules is an important issue to analyze. In addition to these, an added objective lies in the development of power aware scheduling strategies that can ensure prevention from power draining attacks of HTHs.

In present chapter, multi FPGA based systems are considered that have strict power budget. These execute periodic and non-periodic tasks. However, HTHs implanted either in FPGA fabric or in RIPs procured from various 3PIP vendors may cause power draining attacks. We analyze the scenario where HTHs may cause power draining attacks and affect real time task schedules. Such threats may decrease the lifetime of the system and cease its operations before some critical functionalities that are scheduled during the middle or end phases of its expected lifetime. A hybrid online-offline scheduling strategy is proposed to ensure reliable task operations till the system's expected lifetime. In this, self aware agents are generated that detects malfunctioning resources and re-schedules periodic task operations at runtime based on dynamic power dissipation report. The agents also tries to execute maximum incoming non-periodic tasks within pre-specified power budget by allocating them in slack spaces of periodic task schedule. Experimentation is performed with EPFL benchmark tasks for multi FPGA based heterogeneous platforms. Results are depicted with respect to the metrics, task success rate (TSR) for periodic tasks and task rejection rate (TRR) for non-periodic tasks.

This chapter is organized in the following manner. Section 7.2 details the system model, while the threat model is discussed in Sect. 7.3. Section 7.4 is associated with review and limitations of existing power aware strategies for HTH attacks. Self aware security mechanism for real time task schedules from HTH based power draining attacks is presented in Sect. 7.5. Section 7.6 deals with experimentation and results. The chapter concludes in Sect. 7.7.

7.2 System Model

7.2.1 Working of System Components

A system with fn number of FPGAs are considered. The nature of FPGAs may be either same, i.e. homogeneous or different, i.e. heterogeneous. RIPs or bitstreams for configuring the FPGAs to perform different types of operations are to be procured from various 3PIP vendors. Let vn be total number of 3PIP vendors and total bitstreams be b that are supplied by the vendors. As discussed in previous chapters, multiple bitstreams must be procured for a particular task operation to eliminate the issue of duplicate HTHs and ensure diversity. These are stored in system's resource pool or memory.

We consider two task interfaces, via which the tasks arrive: (i) Periodic task interface (PTI) and (ii) Non-periodic task interface (NPTI). Dynamic power dissipation feedback is provided via sensors that depict amount of power dissipated at regular time intervals for task executions on related FPGAs. For the present scenario, the report is received at end of each schedule period.

Other than this, there is a control unit (CU) that is responsible for sending signals for scheduling of tasks on the FPGAs. In normal scenarios, this is done as per the periodic task schedule generated offline for periodic tasks and non-periodic tasks are managed at runtime as per the dynamic power dissipation reports. In addition to this, there exists a task information analyzer (TIA) that is associated with bitstream or RIP selection from the resource pool. Finally, there exists a scheduler that schedules the chosen bitstream by the TIA on the FPGA for executing the particular task. System components is depicted in Fig. 7.1.

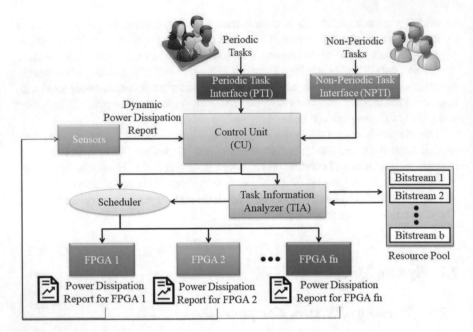

Fig. 7.1 System model

7.2.2   *Nature of Tasks*

Real time periodic tasks and non-periodic tasks have been considered for current study. Periodic and non-periodic tasks are essentially categorized as per their nature of arrival. These were discussed in previous chapters also. We recapitulate them in the current context.

Tasks that arrive or recur after certain regular or fixed time intervals are generally termed as periodic tasks. For periodic tasks, information about task arrival time and task deadline are known in prior, along with the worst case reconfiguration time and execution time for related bitstreams or RIPs of the tasks on different FPGA platforms. Even the approximate power dissipation that will be caused for their execution in various FPGA platforms is even known in advance. As discussed in previous chapters, a periodic schedule can be generated offline, based on which the tasks are scheduled at runtime by the scheduler. However, the offline scheduling in the present case is based on the approximate power dissipation that will be caused. For this present study, it is considered that the periodic tasks are critical in nature and must be completed.

Sporadic and aperiodic tasks for a system are the non-periodic tasks. Their arrival time and deadline is unknown. However, their bitstream reconfiguration and execution time for the various FPGA platforms are known. Even, the approximate values of power dissipation for execution of bitstreams in different FPGAs are available.

The scheduler handles such tasks at runtime. Based on dynamic power dissipation values and remaining power budget, such tasks are scheduled in slack spaces of a periodic task schedule. For present study, non-periodic tasks are considered to be less critical than the periodic tasks.

For task execution on an FPGA platform, dynamic power dissipation is related to various facets like the transistor's effective switching capacity, supply voltage, operating frequency of the FPGA. For offline scheduling, we consider the worst case values of power dissipation for defining the reference power dissipation table. At runtime, supply voltage is kept fixed and issues related to switching capacitance does not affect as we have considered worst case values of power dissipation. Variation of operating frequency of FPGAs are performed in the online phase with the aid of the dynamic clock management (DCM) module. This is done to ensure task completion before deadline during attack scnearios.

7.3 Threat Model

Like the previous chapters, threat associated in the present study also relates to vulnerability in FPGA fabric [MKN+16] and bitstreams or RIPs that are procured from various 3PIP vendors [LRYK14]. The former is associated with HTHs that are essentially malicious circuitry, which are put into the empty spaces of a FPGA layout by adversaries in foundry, during its fabrication. While the latter comprises insertion of malicious codes during hardware description language (HDL) design phase of bitstream generation.

Though the trigger may be the same as discussed in the previous chapters, however, the payload that encapsulates the malicious functional operation is different and comprises of malicious operations that causes unnecessary power dissipation, resulting in reduced lifetime or early expiry of the system. For power draining attacks, the HTH circuitry need not have any direct connection with the original circuit, as illustrated in Fig. 7.2.

The HTH trigger is presented in Fig. 7.2. This is an internal trigger mechanism that comprises a time bomb architecture. This comprises a counter 'x', which increases by one unit with every clock cycle, till a pre-decided value 'y' is reached, which is set by adversary. The malicious functionality of the payload may comprise multiple buffers, or non-functional logic gates, that are not part of the main circuit. There may be loop architectures that enhances the total system's power dissipation, as shown in Fig. 7.2.

If a particular vendor supplies multiple resources, i.e. no diversity in procurement of resources is present, then issue of duplicate HTHs may result. Here, the same HTH may be present in all the procured resources and their trigger times may be the same. In such a scenario, all the HTHs may get activated simultaneously and lead to high power drainage within a short time interval.

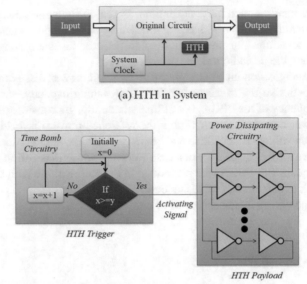

(a) HTH in System

(b) Structure of HTH for Power Draining Attacks

Fig. 7.2   **a** HTH in system **b** structure of HTH for power draining attacks

Table 7.1   Sample periodic task set

Tasks (T_i)	Reconfiguration time (ρ_i)	Execution time (ϵ_i)	Arrival time (α_i)	Deadline (δ_i)	Periodicity (π_i)
T_1	1	6	0	10	20
T_2	2	12	0	18	20
T_3	1	9	5	20	20
T_4	2	5	10	20	20

7.3.1   Illustrative Example

Considering periodic task set of Table 7.1, where task $T_i = \{\rho_i, \epsilon_i, \alpha_i, \delta_i, \pi_i\}$. ρ_i depicts worst case reconfiguration time and ϵ_i represents worst case execution time of related RIPs of T_i. Timings vary with different FPGA platforms and hence, their Worst case timing values are taken into consideration. Task arrival time and deadline are represented by α_i and δ_i respectively. Task periodicity is denoted as π_i, which in the present study is considered to be 20 time units. Four tasks are considered, i.e. $i = \{1, 2, 3, 4\}$. Related task schedule for normal scenario is depicted in Fig. 7.3a. It is to be noted that any type of scheduling mechanism may be considered as per application suitability. However, for simplicity, earliest deadline first is considered in present case. It is also assumed that the operational frequency is f and tasks arrive n times, with a periodicity of 20 time units.

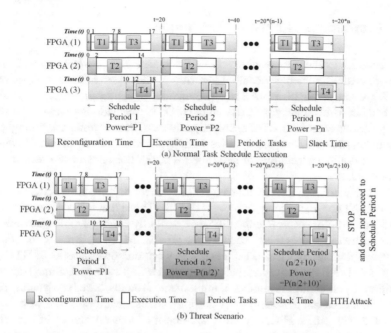

Fig. 7.3 **a** Normal scenario **b** threat scenario

Let us consider that the power dissipation of task T_i is $P(T_i)$. Hence, in a schedule period, effective power dissipation (P_{sp}), will be as Eq. 7.1, where T_{ij} denotes task T_i executing on $FPGA_j$.

$$P_{sp} = \Sigma_i P(T_{ij}) \tag{7.1}$$

P_{sp} denotes summation of power dissipated in a schedule period sp. sp ranges from $\{1, 2, \ldots, n\}$. Thus, total power dissipation is $\Sigma_{sp=1}^{n} P_{sp}$. At present, total power budget of system is considered to be $\Sigma_{sp} P_{sp}$.

Let $n = 400$ and at $n = 200$, HTH attack occurs. Let $FPGA(1)$ and $FPGA(2)$ is associated with a vulnerability. Thus, tasks T_1, T_2 and T_3 will get affected. However, tasks operating on $FPGA(3)$ won't be unaffected, as depicted in Fig. 7.3b. Let $P(T_1)'$, $P(T_2)'$ and $P(T_3)'$ be enhanced power dissipation values. Thus, power dissipation of $sp = n/2$, will be $P(n/2)' = P(T_1)' + P(T_2)' + P(T_3)' + P(T_4)$. If we consider $\Sigma_{sp=(n/2+10)}^{n} P(T_{sp}) = 10 * [P(n/2)' - P(n/2)]$.

Then, total power budget of system is drained in $n/2 + 10$ schedule periods. After that, operations of the system will cease or the system will halt after 210 schedule periods, while it's lifetime is 400 schedule periods. This threat scenario is illustrated in Fig. 7.3b.

7.4   Limitations of Existing Techniques

Recent works in the current domain are associated with development of energy or power aware, efficient scheduling algorithms on trusted hardware. A survey of such is present in [SP18, BMAB16]. A dynamic frequency and scaling based low overhead heuristic strategy is discussed in [MDS19]. Scheduling real time tasks by finding various combinations of energy signatures to deliver optimum throughput is presented in [BC18]. Thus, the works essentially focus on management of power and energy for runtime scheduling of real time tasks, but does not ensure security from hardware attacks.

Works related to reliability driven real time scheduling are yet in infancy stages. Some doesn't consider energy or power [BM13]. Those that take energy and power in consideration for ensuring reliability are essentially confined to normal faults and attacks related to vulnerability of software [Tos12, MFP16]. But do not consider hardware vulnerabilities like HTH attacks. A work on power attacks of HTHs was shown in [MYAH19], however, their aim is to develop various runtime encryption algorithms by changing the number of implemented rounds, based on dynamic power report. This does not consider power draining attacks. Moreover, the work is only associated with single FPGA systems and doesn't focus on multi FPGA systems. HTH attacks for real time schedules and related security mechanisms were also presented in [GSC18, GMSC19, GMSC18], but in all these cases, power draining attacks of HTHs were not focused. Essentially, power draining attacks by HTHs for real time scheduling is still in its nascent stages.

7.5   Self Aware Strategy to Handle Power Draining Attacks

7.5.1   Periodic Task Handling

Offline Task Schedule Generation This phase comprises two steps:
(i) Finding the minimum FPGAs [$min(FPGA)$] and maximum FPGAs [$max(FPGA)$] for all periodic tasks execution. Considering they operate with their default (or lowest) frequency for $min(FPGA)$ and they operate with their maximum permissible (or highest) frequency for $max(FPGA)$. This is presented by Algorithm 1.

(ii) Generation of periodic task schedules for the various FPGAs (that ranges from $min(FPGA)$ to $max(FPGA)$). The mechanism is depicted in Algorithm 2.

For the first step, we consider variable q as an index for the FPGAs. For determining $min(FPGA)$, we initially set $q = 1$. Operational frequency ensuring periodic tasks completion on q resources is assigned as ϕ, which is evaluated as per Eq. 7.2. In this, σ represents total time units in a schedule period, χ denotes time units required for operation of Control Unit, variable i is used to index periodic tasks, ρ_i, ϵ_i depicts reconfiguration and execution time of $T_i.f$ is used for denoting operational frequency. If ϕ is less than $max(f)$ (maximum operable frequency), then value of q is assigned to $min(FPGA)$. Else, q is increased by one unit and ϕ is re-evaluated till it is less than $max(f)$.

$$\phi = \frac{\Sigma_i \epsilon_i}{q(\sigma - \chi) - \Sigma_i \rho_i} \tag{7.2}$$

We start with $min(FPGA) + 1$ for determining $max(FPGA)$. This value is entitled to q. As per Eq. 7.2, ϕ is evaluated and checked if it is less than or equal to $min(f)$. If it is not, then q is incremented by one unit and ϕ is re-evaluated, till $\phi \leq min(f)$. q is assigned to $max(FPGA)$, when $\phi \leq min(f)$.

Algorithm 1: Finding Minimum and Maximum FPGAs Needed for Generation of Periodic Task Schedules

Input: Periodic Task Set
Output: Minimum and Maximum FPGAs Needed
Let $min(FPGA)$ and $max(FPGA)$ denotes minimum and maximum FPGAs needed for scheduling the periodic tasks, variable q index the FPGAs, σ denote the total time units in a schedule period, χ represent time units for operation of Control Unit, i index the periodic tasks, ρ_i represent reconfiguration time of T_i, ϵ_i denote execution time of T_i, f represent default operational frequency and variable ϕ depict the operational frequency needed.
Finding $min(FPGA)$
1. Set $q = 1$;
2. $\phi = \frac{\Sigma_i \epsilon_i}{q(\sigma - \chi) - \Sigma_i \rho_i}$
3. **if** $\lceil \phi \rceil \leq max(f)$ **then**
   $\llcorner$ Set $min(FPGA) = q$;

else
   $\llcorner$ $q = q + 1$ and re-evaluate ϕ or restart from Step 2;
Finding $max(FPGA)$
4. Set $q = min(FPGA) + 1$
5. $\phi = \frac{\Sigma_i \epsilon_i}{q(\sigma - \chi) - \Sigma_i \rho_i}$
6. **if** $\phi <= min(f)$ **then**
   $\llcorner$ Set $max(FPGA) = q$;

else
   $\llcorner$ $q = q + 1$ and re-evaluate ϕ or restart from Step 5;

We assume that there are k different schedules. Let total FPGAs be denoted by fn. Starting with $fn = min(FPGA)$ and $k = 1$, operational frequencies of FPGAs have to be set. Several combinations are possible. As discussed before, q is used as an index for the FPGAs. Let $\phi_q = \mu f$ for $FPGA(q)$. This denotes $FPGA(q)$'s working frequency that is μ times default frequency f.

To determine schedule k or $S(k)$, we begin with first FPGA. Let t represent time instants of $FPGA(q)$. This is set to χ, i.e. control unit's completion time from beginning of schedule period. Next is ordering of available and unscheduled tasks at time t. Tasks with same deadline are ordered on basis of their earliest arrival time. Variable p is used to denote the order and ζ represents total tasks present at t. According to order p, Tasks are fetched and checked. If the task can complete prior to deadline when executed on $FPGA(q)$ at frequency ϕ_q, then the task is marked

---

Algorithm 2: Generation of Schedule Periods for Periodic Task Set

---

Input: Periodic Task Set

Output: k types of Schedules: $S(k)$

1. Set $fn = min(FPGA)$, $k = 1$, where variable fn determines total FPGAs and k indexes possible types of schedules;

2. Set frequencies $\phi_q = \mu f$ for respective $FPGA(q)$, where variable q indexes the FPGAs and q ranges from $\{1, 2, ..., fn\}$ and ϕ_q represents the working frequency of $FPGA(q)$ that is multiple μ of f;

Determining a Schedule $S(k)$

3. Set $q = 1$;

4. **for** $FPGA(q)$, set $t = \chi$, where t represents time instant, while χ represents the time required for the control unit to operate; **do**

    5.Order available and unscheduled tasks at t, first as per their earliest deadline and then as per their earliest arrival time, i.e. tasks that have the same deadline;

    6. **for** $(p = 1; p <= \zeta; p + +)$ *variables p and ζ denote the order and total unscheduled and available tasks at t respectively* **do**

        Fetch T_i at order p

        if $(t + \rho_i + (\lceil \epsilon_i/\phi_q \rceil) < \delta_i$ **then**

            Schedule T_i for execution in $FPGA(q)$ at time t and mark T_i as scheduled;

            Update t, i.e. $t = (t + \rho_i + (\lceil \epsilon_i/\phi_q \rceil)$;

            Repeat from Step 5;

        else

            Return to Step 6;

    if $t < \sigma$, where σ is total time units in a schedule period **then**

        $t = t + 1$ and Repeat from Step 5;

7. **if** $(\zeta \neq 0)||(q < fn)$, *i.e. all tasks not scheduled* **then**

    $q = q + 1$ and Restart from Step 4;

8. Output schedule $S(k)$;

$S(k)$ repeats after every σ time units;

9. **if** *Any other combination of frequency with FPGA is possible for q number of FPGAs* **then**

    Repeat from Step 2;

10. **if** $fn < max(FPGA)$ **then**

    Set $fn = fn + 1$, $k = k + 1$ and Repeat from Step 2;

---

scheduled. Accordingly, t is updated as per its completion time. For new t, the process is repeated from task ordering. On completion of schedule period, if more tasks are left that needs to be scheduled, q is increased by unity and the scheduling mechanism is performed for new $FPGA(q)$. After all tasks are scheduled, i.e. $q == fn$, $S(k)$ is generated as output. With change in operational frequency of FPGAs, various schedules are obtained.

This is shown via the following example.

Illustrative Example Considering periodic task set of Table 7.1 for demonstrating. $\sigma = 20$, $\Sigma_i \rho_i = 6$ and $\Sigma_i \epsilon_i = 32$. Considering $\chi = 3$ and four times possible frequency enhancement is possible, i.e. $max(f) = 4$. Then $\phi = 3$, $min(FPGA) = 1$ when $\phi < max(f)$ and $q = 1$. $\phi = 1.14$ when $q = 2$. Hence, re-evaluation of ϕ is done with $q = 3$. $max(FPGA) = 3$ as $\phi = 0.78$.

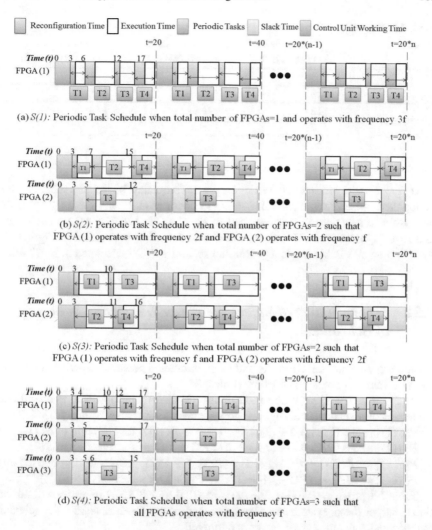

(a) *S(1)*: Periodic Task Schedule when total number of FPGAs=1 and operates with frequency 3f

(b) *S(2)*: Periodic Task Schedule when total number of FPGAs=2 such that
FPGA (1) operates with frequency 2f and FPGA (2) operates with frequency f

(c) *S(3)*: Periodic Task Schedule when total number of FPGAs=2 such that
FPGA (1) operates with frequency f and FPGA (2) operates with frequency 2f

(d) *S(4)*: Periodic Task Schedule when total number of FPGAs=3 such that
all FPGAs operates with frequency f

Fig. 7.4 a Schedule $S(1)$ where there is a single FPGA ($fn = 1$) that operates at $3f$ **b** schedule $S(2)$ where there are two FPGAs ($fn = 2$), $FPGA(1)$ operates at $2f$ and $FPGA(2)$ operates at f **c** schedule $S(3)$ where there are two FPGAs ($fn = 2$), $FPGA(1)$ operates at f and $FPGA(2)$ operates at $2f$ **d** Schedule $S(4)$ where there are three FPGAs ($fn = 3$) and each operates at f

Table 7.2 Power dissipation of tasks

Tasks	Power units for FPGA(1) ($P(T_{i1})$)	Power units for FPGA(2) ($P(T_{i2})$)	Power units for FPGA(3) ($P(T_{i3})$)
T_1	5	7	9
T_2	18	21	25
T_3	12	14	17
T_4	7	9	11

If $fn = 1$, there is a single FPGA operating at frequency $3f$. Related schedule $S(1)$ is depicted in Fig. 7.4a. If $fn = 2$, $FPGA(1)$ operates at $2f$ and $FPGA(2)$ operates at f. Related schedule $S(2)$ is depicted in Fig. 7.4b. But if $FPGA(1)$ operates at f and $FPGA(2)$ operates at $2f$, then related schedule $S(3)$ is depicted in Fig. 7.4c. If $fn = 3$, then all FPGAs work at f. Related schedule $S(4)$ is depicted in Fig. 7.4d.

7.5.2 Determination of Reference Power Dissipation Values of Schedules

Assuming, there are three FPGAs and their associated power dissipation for related tasks of Table 7.1 are as provided in Table 7.2.

As discussed previously, power dissipation is directly proportional to operational frequency. For schedules of Fig. 7.4, power dissipation for a schedule period are:

$P[S(1)] = 3 * [P(T_{11}) + P(T_{21}) + P(T_{31}) + P(T_{41})]$ for selection of $S(1)$,

$P[S(2)] = 2 * [P(T_{11}) + P(T_{21}) + P(T_{41})] + P(T_{32})$ for selection of $S(2)$,

$P[S(3)] = [P(T_{11}) + P(T_{31})] + 2 * [P(T_{22}) + P(T_{42})]$ for selection of $S(3)$ and

$P[S(4)] = P(T_{11}) + P(T_{22}) + P(T_{33}) + P(T_{41})$ for selection of S(4).

Table 7.3 provides $P[S(1)]$, $P[S(2)]$, $P[S(3)]$ and $P[S(4)]$ of various combinations. As per these, system architecture resources are determined and the schedule that dissipates the minimum power is deployed.

For the current study, considering tasks of Table 7.1, three FPGAs are considered and $S(4)$—*Option*1 followed, as related power dissipation is minimum. But if there is constraint in the number of resources and if one FPGA is deployed, then $S(1)$—*Option*1 should be considered.

However, it is to be noted that security is of prime concern for critical infrastructures. Hence, maximum power budget is to be deployed with high processing resources. This is necessary as for attack scenarios, additional power can be used for re-execution of affected tasks in non-affected resources. If no threat occurs, then the extra resources and power can be utilized for execution of non-periodic tasks.

Table 7.3 Schedule power dissipation values

Options	$P[S(k)]$(power units for $S(k)$)	Comments
On deployment of $S(1)$		
1	126	In $FPGA(1)$
2	153	In $FPGA(2)$
3	186	In $FPGA(3)$
On deployment of $S(2)$		
1	74	$FPGA(1)$ frequency: $2f$; $FPGA(2)$ frequency: f
2	77	$FPGA(1)$ frequency: $2f$; $FPGA(3)$ frequency: f
3	91	$FPGA(2)$ frequency: $2f$; $FPGA(3)$ frequency: f
4	86	$FPGA(2)$ frequency: $2f$; $FPGA(1)$ frequency: f
5	102	$FPGA(3)$ frequency: $2f$; $FPGA(1)$ frequency: f
6	104	$FPGA(3)$ frequency: $2f$; $FPGA(2)$ frequency: f
On deployment of $S(3)$		
1	77	$FPGA(1)$ frequency: f; $FPGA(2)$ frequency: $2f$
2	87	$FPGA(1)$ frequency: f; $FPGA(3)$ frequency: $2f$
3	93	$FPGA(2)$ frequency: f; $FPGA(3)$ frequency: $2f$
4	71	$FPGA(2)$ frequency: f; $FPGA(1)$ frequency: $2f$
5	76	$FPGA(3)$ frequency: f; $FPGA(1)$ frequency: $2f$
6	86	$FPGA(3)$ frequency: f; $FPGA(2)$ frequency: $2f$
When schedule $S(4)$ is deployed		
1	50	Order: $FPGA(1)$; $FPGA(2)$; $FPGA(3)$
2	89	Order: $FPGA(1)$; $FPGA(3)$; $FPGA(2)$
3	53	Order: $FPGA(2)$; $FPGA(3)$; $FPGA(1)$
4	51	Order: $FPGA(2)$; $FPGA(1)$; $FPGA(3)$
5	52	Order: $FPGA(3)$; $FPGA(1)$; $FPGA(2)$
6	53	Order: $FPGA(3)$; $FPGA(2)$; $FPGA(1)$

7.5.3 Mechanism for Detection of Affected Resources

As previously discussed, threat in the present scenario is either due to vulnerabilities in the procured RIPs or in the FPGA fabric. In addition to this, possibility remains in the existence of duplicate HTHs [GSC19b], if the resources are procured from the same vendor. Thus, to eradicate the issue of duplicate HTHs, it is best to procure each resource from a different vendor. But this will add in enhanced license fees and cost of operation.

RIPs are essentially HDL files that are of a few kilobytes in size, while the FPGAs define the physical resource of the system, where actual task execution takes place. While the FPGAs are costly, the RIPs are comparatively cheap. Moreover, for a system architecture, only few FPGAs are used, while numerous RIPs need to be stored for the various types of tasks that are intended to be executed. Hence, it is wise to procure each FPGA from a different source, so that issues related to diversity do not matter. While multiple RIPs may be procured from a 3PIP vendor, but for a particular operation at least three RIPs from three different vendors need to be procured to ensure diversity, as discussed in previous chapter.

In the current study, we consider development of a self aware agent (SAA) that works based on the Observe-Decide-Act philosophy of previous chapters. The SAA may be either a part of the control unit or a separate entity that possess the ability to direct the operations of the control unit. For the current study, we consider the SAA logic to be embedded in the control unit of the system. Detection of malware or HTHs is performed based on dynamic power dissipation values received as feedback from the sensors. Dynamic power dissipation values are considered for detection of HTHs as the considered attacks neither generates wrong results, nor causes timing issues, but are associated with power draining or excessive power dissipation that leads to early expiry of power budget of the system.

Detection of malware is performed based on five rules:

(i) **Rule 1**: No system vulnerability if power dissipation of each task is less or equal to reference values.

In such a scenario, no malicious actions occur in either the RIPs or the FPGAs and hence, normal operations can take place.

(ii) **Rule 2**: FPGA is affected if all tasks that execute in an FPGA exceed their power dissipation limits.

This is because if the FPGA is associated with an HTH, then all tasks that operates will be affected. In such a case, the FPGA needs to be isolated and no operations are to take place on it in future.

Note: This rule is not valid if a single task operates in an FPGA.

(iii) **Rule 3**: If multiple tasks executing in different FPGAs are associated with increased power dissipation, and RIPs of all such tasks are procured from a particular vendor, then the 3PIP vendor is malicious.

This is a case of implantation of duplicate HTHs by the 3PIP vendor, where all such HTHs implanted in different RIPs activates at the same time.

For this scenario, the 3PIP vendor needs to be blacklisted, so that none of the RIPs procured from it is used in future operations.

(iv) **Rule 4**: If multiple task operations are associated to an FPGA, but all tasks does not exhibit increase in power dissipation, then 3PIP vendors that supplied RIPs are malicious.

As if HTHs is present in FPGA fabric, then all tasks will endure enhanced power dissipation.

Remedy must be taken as per Rule 3.

(v) **Rule 5**: If only a single task executes on an FPGA, which is associated with increased power dissipation, then re-evaluation have to be done.

As either FPGA fabric or RIP can be affected. In re-evaluation, a different RIP is to be executed on the same FPGA platform. If enhanced power dissipation is observed in the new case, then the FPGA is affected (due to similar reasons, as per Rule 2), and have to be isolated. Else the RIP is affected and suitable actions are to be taken as per Rule 3.

Note 1: For multiple task executions that take place on a single FPGA of a schedule period, RIPs from at least two different vendors must be utilized.

Note 2: Similar HTH architectures cannot be implanted in resources procured from different sources. In such cases, HTH trigger timing will definitely be different, though their payloads may be same [RSK16, CMSW14].

Note 3: For cases where power dissipation of individual tasks is not available and total power dissipation of the schedule is obtained (as different tools provide reports in different ways), fault diagnosis is to take place in the next schedule period. In fault diagnosis, each task of the schedule is executed separately on the FPGA platform and their individual power dissipation values are sent as feedback for analysis.

Algorithm 3 depicts the mechanism, which is even discussed as follows:

For detecting affected resources, a dedicated counter is deployed for each FPGA, which is denoted by $COUNT_{FPGA(q)}$ and a dedicated counter is also deployed for each 3PIP vendor, designated as $COUNT_{Vendor(v)}$. Out of order task executions, i.e. task executions that are associated with enhanced power distributions for $FPGA(q)$ is counted by $COUNT_{FPGA(q)}$. Similarly, $COUNT_{Vendor(v)}$ counts out of order task executions related to RIPs supplied by $Vendor(v)$, where vn is the number of 3PIP vendors and range of v is from $\{1, 2, \ldots, vn\}$.

Let q index the FPGAs. For $S(k)$, we begin with $q = 1$ or the first FPGA. For each task, $(P_{observed}T_{iq})$ (actual power dissipation obtained from sensor report) is compared with $(P_{reference}T_{iq})$ (reference power dissipation), for a particular operational frequency ϕ_q. If equivalent, no malicious activity is present. Else both $COUNT_{FPGA(q)}$ and $COUNT_{Vendor(v)}$ is enhanced by unity, where $Vendor(v)$ refer to the 3PIP vendor that supplied the RIP for the particular operation. This is performed for all task operations related to $S(k)$.

If $FPGA(q)$ count, i.e. $COUNT_{FPGA(q)}$ is 0, then FPGA can be marked as safe (Rule 1). However, if multiple tasks operate on $FPGA(q)$, but none are affected, then $FPGA(q)$ can also be marked safe (Rule 4). But if $COUNT_{FPGA(q)}$ is equal to its total task executions, which is denoted by iqn, and $iqn > 1$, then the FPGA can be considered affected and is to be marked unsafe (Rule 2). For the case of single

---

Algorithm 3: Detecting Affected Resources

Input: Periodic Task Schedule $S(k)$

Output: Status of Resources

For detecting the affected resources and malicious vendors, we use several counters, where variable $COUNT_{FPGA(q)}$ indicate a count of malicious task executions in $FPGA(q)$ and variable $COUNT_{Vendor(v)}$ indicate count of malicious tasks associated with RIPs procured from $Vendor(v)$;

For re-evaluation in next schedule period, variables $CHECK(q)$ are used.

for $(q = 1; q <= fn; q + +)$ *variables fn denote the total number of FPGAs for input schedule $S(k)$ and q is a variable for indexing the FPGAs* **do**

    for $(iq = 1; iq <= iqn; iq + +)$ *variables iqn represent the total tasks for execution on FPGA(q) working with a frequency of ϕ_q in schedule $S(k)$ and iq is a variable for indexing* **do**

        if $[P_{observed}(T_{iq})/\phi_q] > [P_{reference}(T_{iq})/\phi_q]$ **then**

            $COUNT_{FPGA(q)} = COUNT_{FPGA(q)} + 1$;

            for $(v = 1; v <= vn; v + +)$, *where variables vn represent the total number of vendors from which bitstreams for task execution are procured and v is a variable for indexing the total number of vendors* **do**

                if $Bitstream(T_{iq})$ *is procured from 3PIP Vendor(v)* **then**

                    $COUNT_{Vendor(v)} = COUNT_{Vendor(v)} + 1$

    if $COUNT_{FPGA(q)} == 0$ **then**

        Mark $FPGA(q)$ as SAFE *(RULE 1)*;

    if $(COUNT_{FPGA(q)} < iqn)$ **then**

        Mark $FPGA(q)$ as SAFE *(RULE 4)*;

    if $(COUNT_{FPGA(q)} == iqn)\&\&(iqn > 1)$ **then**

        Mark $FPGA(q)$ as UNSAFE *(RULE 2)*;

    if $((COUNT_{FPGA(q)} == iqn)\&\&(iqn == 1)\&\&(CHECK(q) == 0))$ **then**

        Set $CHECK(q) = 1$ *(RULE 5)*;

    if $((COUNT_{FPGA(q)} == iqn)\&\&(iqn == 1)\&\&(CHECK(q) == 1))$ **then**

        Mark $FPGA(q)$ as UNSAFE *(RULE 5)*;

    else

        Mark $FPGA(q)$ as SAFE and set $CHECK(q) = 0$ *(RULE 5)*;

for $(v = 1; v <= vn; v + +)$ **do**

    if $COUNT_{Vendor(v)} == 0$ **then**

        Mark $Vendor(v)$ as SAFE *(RULE 1)*;

    else if $(COUNT_{Vendor(v)} > 0)$ *&& (All FPGA(q) are SAFE)* **then**

        Mark $Vendor(v)$ as UNSAFE *(RULE 3 and 4)*;

---

task execution on an FPGA, i.e. $iqn == 1$, then either FPGA or RIP can be affected. For this, a $CHECK(q)$ variable is utilized that is initially set to 1. Re-evaluation is to be performed when $CHECK(q) = 1$. For re-evaluation, in next schedule period, a different RIP procured from an alternate vendor but performing the same operation is deployed for task execution on $FPGA(q)$. If power dissipation exceeds the reference value for re-evaluation, then $FPGA(q)$ is affected, else RIP is affected (Rule 5).

After verification of FPGAs, RIPs are to be checked. Analysis of $COUNT_{Vendor(v)}$ is performed for each vendor v. If $COUNT_{Vendor(v)}$ is 0, then no HTH is implanted and RIPs from that vendor can be considered safe for future operations (Rule 1). Else if $COUNT_{Vendor(v)} > 0$ and all FPGAs are marked safe, then vulnerability exists in RIPs supplied by that vendor. In such scenario, the vendor is to be blacklisted and all RIPs procured from that vendor are to be ignored for future operations (Rule 3, Rule 4).

Illustrative Example Considering schedule of Fig. 7.4d and existence of two vendors, $Vendor(1)$ and $Vendor(2)$, we demonstrate the scenario. We also assume that $Vendor(1)$ supply RIPs for T_1 and T_2, while $Vendor(2)$ supply RIPs for T_3 and T_4. If power dissipation of both T_1 and T_4 is greater than their reference values, then $FPGA(1)$ is marked $UNSAFE$. This is because possibility of implantation of same HTH in RIPs procured from different sources is negligible. But if power dissipation of T_1 is higher than its reference, but not of T_4, then $FPGA(1)$ is safe but $Vendor(1)$ is not. $Vendor(1)$ is also unsafe if power dissipation of T_1 and T_2 is higher than their reference values.

If there is an increase of power in T_3 executing $FPGA(3)$, then it can be either due to HTH present in $FPGA(3)$ or HTH in RIP supplied by $Vendor(2)$. $CHECK(3)$ is set to 1 for this case. In subsequent schedule period, a different RIP is used for operation of T_3 on $FPGA(3)$ (may be from $Vendor(1)$ or a new vendor, $Vendor(3)$). If enhanced power dissipation continues, then $FPGA(3)$ is unsafe, else $Vendor(1)$ is unsafe and all RIPs procured from $Vendor(1)$ is to be ignored in future.

7.5.4   Action on Detection of Vulnerability

On detection of vulnerability, the SAA isolates the affected resources and tries to execute the critical tasks in non-affected resources. For the present study, periodic tasks are considered critical in nature and hence, these are given preference. The objective is to ensure execution of pre-scheduled periodic tasks in non-affected resources till system lifetime.

Deployment mechanism by SAAs in safe resources is described in Algorithm 4.

Let at a certain time instant, only s number of safe FPGAs are present. Thus, a schedule of s FPGAs is selected. Let the FPGAs are indexed by variable q, vn denote the number of 3PIP vendors and let variable v index vendor based RIP allocation scheme. We consider that all vendors are capable of supplying RIPs for all task operations. Now, for each task, it is seen whether $CHECK(q)$ for $FPGA(q)$ is set to 1 or not. Normal is the scenario if $CHECK(q) = 0$. Then it is seen whether $Vendor(v)$ is safe or not. If not, then $Vendor(v + 1)$ (i.e. vendor next in order) is selected. The process continues until a safe vendor is selected. RIPs supplied by that safe vendor is allocated for execution. But if $CHECK(q) = 1$, then RIP from another vendor have to be assigned. For this mechanism, v is incremented by one, and checked till a safe vendor is found. It may occur that v exceeds vn (i.e. total vendors). To tackle such a scenario, a modulus operation is deployed. Via this, after $v = vn$, v restarts from 1. Each FPGA's operational frequency is set as per selected schedule.

Algorithm 4: Action Mechanism on Detection of Discrepancy

Input: Periodic Task Schedules, Status of Resources
Output: Schedule Deployment
Let s denote the number of SAFE FPGAs, fn denotes the total FPGAs in system and vn be
the total number of vendors from whom RIPs are obtained;
Set $v = 1$, where variable v is used for indexing the number of vendors;
Select a periodic task schedule with s number of FPGAs;
Mechanism of Deployment of Schedule
1. **for** $(q = 1; q <= fn; q++)$ *where q is a variable for indexing the FPGAs* **do**
    if *FPGA(q) is SAFE* **then**
        2. **for** $(iq = 1; iq <= iqn; iq++)$ *variables iqn represent the total tasks for*
        execution on FPGA(q) working with a frequency of ϕf and iq is a variable for
        indexing **do**
            3. **if** $CHECK(q) == 0$ **then**
                4. **if** *Vendor(v(mod vn)) is SAFE* **then**
                    ⌊ Deploy *Bitstream(Tiq)* procured from *Vendor(v(mod vn))*

                else
                    ⌊ $v = v + 1$ and re-perform Step 4;

            5. **if** $CHECK(q) == 1$ **then**
                6. $v = v + 1$;
                7. **if** *Vendor(v(mod vn)) is SAFE* **then**
                    ⌊ Deploy *Bitstream(Tiq)* procured from *Vendor(v(mod vn))*

                else
                    ⌊ $v = v + 1$ and re-perform Step 7;

        8. Set frequency of operation of *FPGA(q)* to ϕf as per chosen schedule;

Illustrative Example For demonstration, tasks of Table 7.1 and schedules of Fig. 7.4 are considered. Handling FPGA attacks is discussed in Case 1, while handling attacks on RIPs is discussed in Case 2.

Scenario 1: Handling Attacks related to FPGAs
First, let us choose $S(4)$ and consider $n = 400$. We assume $FPGA(1)$ is implanted with a HTH, which triggers at $n = 200$. This causes power of T_1 and T_4 to increase. Thus, a two FPGA based schedule is to be considered, i.e. $S(2)$ or $S(3)$. Probability of $S(3)$ is more as its total power is less than $S(2)$. As no issues are associated with vendors, so, their ordering is unimportant. It is possible that after certain time span, another FPGA exhibits malicious behavior. Let $S(3)$ be deployed and power dissipation values of tasks T_2 and T_4 exhibit an enhancement at $n = 275$. Then it can be considered that $FPGA(3)$ is also affected. Then, $S(1)$ have to be deployed. Figure 7.5a depicts this scenario.

If $S(2)$ is chosen and at $n = 275$, it is seen that power of T_3 has increased. Then either $FPGA(3)$, or RIP of T_3 may be affected. For this, $CHECK(3)$ is set to 1, at $n = 276$. T_3 is re-executed in $FPGA(3)$, but with RIP from a different 3PIP vendor. If enhanced power is still observed than reference, then $FPGA(3)$ is malfunctioning. From next schedule period, $S(1)$ is to be deployed in $FPGA(2)$, as shown Fig. 7.5b.

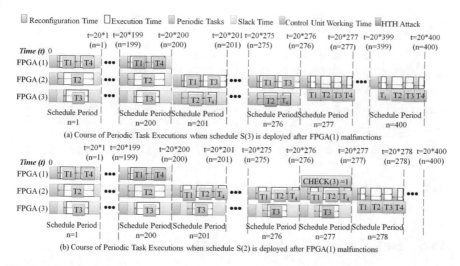

Fig. 7.5 Periodic task executions for 'n' schedule periods, where 'n = 400', when FPGAs are affected

But if after a certain stage, the last FPGA also malfunctions, then system stops. Such a probability is quite low, where all FPGAs procured from different sources will be associated with malware.

Case 2: Handling Attacks related to RIPs

We assume that there are four vendors and all can supply RIPs related to various task executions. Periodic task schedule as depicted in Fig. 7.4d is considered for demonstration. Let RIP for T_1, T_2, T_3, T_4 is procured from $Vendor(1)$, $Vendor(2)$, $Vendor(3)$, $Vendor(4)$, respectively, These are deployed in all schedule periods.

Let us consider a threat where an HTH is present in an RIP, procured from $Vendor(2)$. It triggers at $n = 200$. Thus, enhanced power dissipation is observed for T_4, and $Vendor(2)$ is marked UNSAFE. For $n = 201$, T_1, T_4, T_2 from $Vendor(1)$, $Vendor(3)$, $Vendor(4)$ will be used respectively, as per Algorithm 4, till further anomalies is detected. Let enhanced power dissipation is seen in T_2 at $n = 276$. Then, either $FPGA(2)$ or RIP from $Vendor(4)$ is associated with a vulnerability. Then, $CHECK(2)$ is set to 1. From subsequent schedule period, RIP for T_1 is procured from $Vendor(1)$, T_4 is procured from $Vendor(3)$, T_2 is procured from $Vendor(1)$ instead of $Vendor(4)$ and T_3 is procured from $Vendor(3)$, as per Algorithm 3. If no enhancement in power dissipation is seen in T_2 when operated with RIP from $Vendor(1)$, then $FPGA(2)$ is not malicious and $Vendor(4)$ is unsafe. From $n = 278$, as per Algorithm 4, RIP for T_1 is procured from $Vendor(1)$, T_4 is procured from $Vendor(3)$, T_2 is procured from $Vendor(1)$ and T_3 is procured from $Vendor(3)$, till further malicious operation is detected. This is shown in Fig. 7.6.

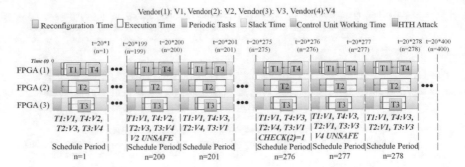

Fig. 7.6 Periodic task executions for 'n' schedule periods, where 'n=400', when RIPs are affected

7.5.5 Handling of Non-periodic Tasks

Non-periodic tasks that arrive via the non-periodic task interface (NPTI) are managed at runtime by the control unit. The mechanism is presented via Algorithm 5.

Let *npn* denote total non periodic tasks and *s* represent total number of safe FPGAs at a certain time instant. Thus, periodic task schedule that requires *s* FPGAs is chosen, in which the objective lies in accommodation of maximum amount of non-periodic tasks. Variables t, np, q, ϕ_q denote the time instant, non-periodic tasks, variable for indexing of FPGAs and operational frequency of $FPGA_q$ respectively. Let for an available period, its start time be represented by $Start(AP)$ and end time by $End(AP)$, where a non-periodic task is expected to be executed. Available period or suitable time slack for each non-periodic task is searched in each FPGAs of the selected schedule. $Start(AP)$ and $End(AP)$ are initially 0. If at t, a slack space is present, then $Start(AP)$ is set to t. If not, t is increased by 1, unless suitable slack is detected. Now, searching is done from $Start(AP)$ to δ_i (task deadline), to find a suitable available period. For this, t is increased till task deadline or no more slack is present. $End(AP)$ represents this time instant. If task execution can complete before deadline in the available period, i.e. summation of reconfiguration and execution time of related RIP on $FPGA_q$, at operational frequency ϕ_q does not exceed available period time and additional power that will be consumed for that task execution (i.e. $(P(T_i) + P_{sp}reference)$) will not be more than the schedule period's total power, then non-periodic task's execution is scheduled in the available period. Else process is repeated by enhancing the value of t by 1.

Illustrative Example Let at $sp = 2$, arrival of two non-periodic tasks take place as detailed in Table 7.4, whose reference power values is detailed in Table 7.5. These needs to be allocated in slack spaces of periodic task schedules of Fig. 7.4.

For S(1), neither T_5, nor T_6 can be accommodated as slack is not available.

Let's consider S(2) and S(3). T_5 is rejected, as no suitable slack is present in that time frame or no available period can be found. But for T_6, available period from time $t = 12$ to time $t = 20$ in $FPGA_2$ of S(2) and from time $t = 16$ to time

---

Algorithm 5: Non-Periodic Tasks Handling

Input: Deployed Periodic Task Schedule
Output: Schedule with Non-Periodic Tasks for Current Schedule Period
Let total non-periodic tasks be represented by npn and s denote total SAFE FPGAs, while variables np and q are used for indexing non-periodic tasks and FPGAs respectively. ϕ_q represents the working frequency of $FPGA_q$;
For the periodic task schedule with qn FPGAs;
1. **for** *(np=1 to npn)* **do**
    2. **for** *(q=1 to qn)* **do**
       3. Set Start(AP)=0, End(AP)=0
       It is assumed that variable t is used for indexing the time and Start(AP) and End(AP)
       variables are used for denoting an available period
       4. **if** *Slack present at t* **then**
         ⌊ Set Start(AP)=t;
       else
         ⌊ Increment t, i.e. t = t + 1 and Redo Step 4;
       5. **for** $t = Start(AP)$ to δ_i **do**
         if $t == \delta_i$ **then**
           ⌊ Set End(AP) $= \delta_i$;
         else if *Slack present at t* **then**
           ⌊ Increment t, i.e. t = t + 1 and Repeat Step 5;
         else
           ⌊ Set End(AP) = t;
       6. **if** $(Start(AP) + \rho_i + \epsilon_i/\phi_q) < (End(AP)) \&\&$
       $((P(T_i) + P_{sp}reference) < P_{sp}available)$ **then**
         ⌊ Schedule task T_i in $FPGA_q$ from Start(AP) to End(AP)
       else
         ⌊ Increment t, i.e. t = t + 1 and Repeat from Step 3;

Table 7.4 Non-periodic tasks

Tasks (T_i)	Reconfiguration time (ρ_i)	Execution time (ϵ_i)	Arrival time (α_i)	Deadline (δ_i)
T_5	1	2	5	10
T_6	1	4	10	20

Table 7.5 Power dissipation values of non-periodic tasks

Tasks	Power units in FPGA(1) ($P(T_{i1})$)	Power units in FPGA(2) ($P(T_{i2})$)	Power units in FPGA(3) ($P(T_{i3})$)
T_5	3	5	8
T_6	4	7	10

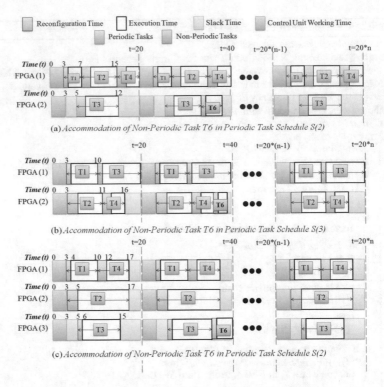

Fig. 7.7 Accommodation of non-periodic task T_6 in **a** S(2) **b** S(3) **c** S(4)

$t = 20$ in $FPGA_2$ of S(3) can be found. Such available spaces are sufficient for reconfiguration and execution of RIPs as shown in Fig. 7.7a, b respectively. In S(2), operating with default frequency f, T_6 completes at $t = 17$, while T_6 operating with enhanced frequency $2f$, can complete at $t = 19$ in S(3). Before deploying, power analysis is to be performed. If 75 units is the total available power for S(2):Option 1, then T_6 is not acceptable as it will need 5 additional power units. Similarly, it follows for the other options of schedule S(2) and schedule S(3).

For S(4), T_5 is rejected due to unavailability of slack spaces or available period. Though $FPGA_1$ and $FPGA_2$ have available periods for T_6, but this time is not enough for reconfiguration and execution of RIPs. But for $FPGA_3$, available period is suitable for T_6, as denoted in Fig. 7.7c. If power requirement satisfies, task can be scheduled for operation.

7.6 Experimentation and Result Analysis

7.6.1 Experimentation

Heterogeneous multiple FPGAs are used for ZynQ, Virtex 7 and Spartan 6 FPGAs. Control unit logic is implanted in a soft core processor. As discussed previously, the SAA logic is also embedded as a part of the contol unit. Tasks from EPFL benchmark suite is used for experimentation [AGDM15]. Related RIPs are designed in Verilog and synthesized in Xilinx Vivado platform.

Threat generation is done as per the previous chapters. For RIP related threats, HTH code is implanted during HDL design phase of RIP generation. For FPGA related threats, a small virtual portion is created in the FPGA fabric, where HTH logic is inserted. Structure of HTH used for experimentation is as follows. The HTH trigger comprises five bit counter. When it reaches '11111', HTH payload is activated. Payload architecture comprises a loop architecture with unnecessary gates that dissipates power.

Power Consumption Tables Tables 7.6, 7.7 and 7.8 provides reference power dissipation values of RIPs before (pre-trigger) and after (post-trigger) of HTH activation, for various tasks in different FPGA platforms. Xilinx XPower Analyzer is used for power analysis. As evident, power dissipation values is quite negligible when HTH is dormant or not activated. So, HTH detection is difficult before its trigger. But after activation, power consumption is noteworthy. Various schedules are considered with different tasks by varying their deadline and arrival time. Accordingly, schedule period length varies. For experimentation, it varies from 100 to 500 ns, while number of schedule periods is fixed at 100. As discussed, periodic schedules are generated before deployment, while non-periodic tasks are managed at runtime as per FCFS (first come first serve) policy by the control unit.

Table 7.6 Power dissipation table for $FPGA_1$, i.e. ZYNQ

EPFL bench-marks	Priority encoder	Decoder	Adder	Lookahead XY router	i2c con-troller	Log2	Square root	Coding cavlc
Power (mW) (pre HTH trigger)	2.989	0.412	0.623	0.699	1.302	3.149	4.336	2.151
Power (mW) (post HTH trigger)	4.531	0.673	0.931	1.036	1.876	4.813	6.924	3.033

Table 7.7 Power dissipation table for $FPGA_2$, i.e. Spartan 6

EPFL benchmarks	Priority encoder	Decoder	Adder	Lookahead XY router	i2c controller	Log2	Square root	Coding cavlc
Power (mW) (pre HTH trigger)	3.183	0.517	0.502	0.444	1.454	3.009	4.945	2.140
Power (mW) (post HTH trigger)	4.767	0.199	0.786	0.624	2.190	4.509	7.878	3.314

Table 7.8 Power dissipation table for $FPGA_3$, i.e. Virtex 7

EPFL benchmarks	Priority encoder	Decoder	Adder	Lookahead XY router	i2c controller	Log2	Square root	Coding cavlc
Power (mW) (pre HTH trigger)	2.889	0.708	0.669	0.899	1.239	3.854	4.438	2.148
Power (mW) (post HTH trigger)	4.335	1.132	1.074	1.337	1.753	5.781	6.222	3.436

7.6.2 Result Analysis for Periodic Tasks

For analyzing results of periodic tasks, a metric Task Success Rate (TSR) is defined. TSR is the ratio of total periodic tasks completed successfully over total periodic tasks in a schedule, that are subjected to HTH attacks. This is expressed as a percentage, as shown in Eq. 7.3.

$$TSR = \frac{No.\ of\ Periodic\ Tasks\ Successful}{Total\ no.\ of\ Periodic\ Tasks\ in\ a\ Schedule} * 100\% \qquad (7.3)$$

TSR Analysis over Normalized Power Budget (NPB) and Normalized Time of Attack (NToA) Various periodic schedules are generated for experimentation, where schedule length varies due to varying number of tasks. Normalized values are considered for attaining uniformity during analysis.

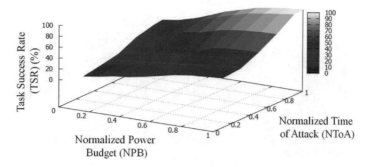

Fig. 7.8 TSR analysis against NPB and NToA

NPB is the ratio of power budget of individual task schedules to maximum power budget among all schedules. Equation 7.4 defines NPB.

$$NPB = \frac{Power\ Budget\ for\ Individual\ Schedule}{Maximum\ Power\ Budget\ among\ all\ Schedules} \qquad (7.4)$$

NToA is the ratio of time of attack on an individual schedule to maximum schedule length among all task schedules. Equation 7.5 defines NToA.

$$NToA = \frac{Time\ of\ Attack\ on\ an\ Individual\ Schedule}{Maximum\ Schedule\ Length\ among\ all\ Schedules} \qquad (7.5)$$

Figure 7.8 shows how TSR varies against NPB and NToA.

TSR increases when NPB and NToA increases. But increase of TSR is slow in the begining, then steadily increases during mid phases and stabilizes at end. This is because in the begining, task schedules are densely packed without any slacks but afterwards, they are loosely packed with slacks and at end, huge slacks are available. With slacks, fault diagnosis is easier and thus, fault detection and recovery can also be performed easily.

Additionally, it can be seen that with low power budget and early attacks, TSR is worst. While with high power budget and late attacks, TSR provides best results. But TSR can never be 100%, as an HTH attack at end of a schedule may not provide ample time for fault diagnosis and recovery. However, effect is negligible when system is near expiry.

TSR Analysis over NToA and FPGA Frequency Fig. 7.9 depicts how TSR varies against increase in FPGA frequency and NToA.

Power consumption increases with increase in operational frequency of an FPGA. Hence, for experimentation, we we scale the operational frequency upto 2^2 times times, though increment upto 2^5 times is possible.

As evident, TSR increases and reaches its peak with increase in operational frequency, which at present is scaled to 2^2 times, except when attack takes place quite early.

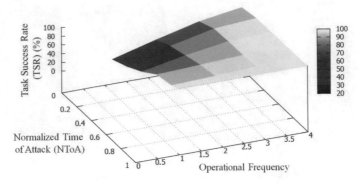

Fig. 7.9 TSR analysis against NToA and frequency of FPGAs

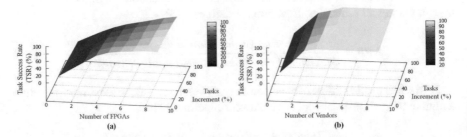

Fig. 7.10 TSR analysis with increase in tasks with respect to **a** increase in FPGAs **b** increase in vendors that supplies RIPs or bitstreams

TSR Analysis over Number of FPGAs and Number of Vendors Fig. 7.10a depicts how TSR varies against increase in FPGAs and increase or percentage of tasks from 0–100%. This aids in determining the maximum number of FPGAs required to achieve a significant TSR. From the analysis, it is evident that when FPGAs increase, TSR increases significantly and reaches a steady state when the number of FPGAs is eight. Moreover, with increase in task percentage, more FPGAs are required for executing them.

Figure 7.10b depicts variation of TSR with increase of 3PIP vendors and percentage increment of tasks from 0–100%. This analysis is important in determining the total 3PIP vendors required that supplies RIPs. Thus, it depicts the diversity needed in RIP procurement. As evident from the analysis, increase in TSR is quite steep and saturation is reached in early stages. Thus, at least two RIPs from two different vendors is needed for the present case. But if simultaneous faults occur, then at least one additional vendor is to be used. However, a maximum of four vendors are required as per the analysis.

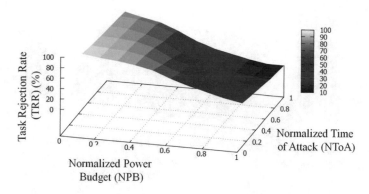

Fig. 7.11 TRR analysis against NPB and NToA

7.6.3 Result Analysis for Non-periodic Tasks

Task Rejection Rate (TRR) is basically a ratio of non-periodic tasks rejected over total non-periodic tasks that arrive in a schedule. This is expressed as a percentage, as shown in Eq. 7.6. TRR is used for analysis of non-periodic tasks.

$$TSR = \frac{No.\ of\ Non\text{-}Periodic\ Tasks\ Rejected}{Total\ no.\ of\ Non\text{-}Periodic\ Tasks\ arrived\ in\ a\ Schedule} * 100\% \quad (7.6)$$

TRR Analysis over NPB and NToA Fig. 7.11 shows how TRR varies against NPB and NToA.

As evident, when attack takes place late and there is an increase in power budget, the TRR decreases. As the schedules are tightly packed in beginning with little or no slack spaces, hence TRR is quite high but as time increases, slack spaces also increase, which in turn decreases TRR, as the non-periodic tasks can be allocated for execution in the available slacks. Finally at end of schedule, TRR stabilizes as huge slacks are available at end phases.

TRR Analysis over Normalized Task Relative Deadline (NTRD) with respect to NPB Relative deadline of a non-periodic task can be stated as time period from the task's time of arrival to the task's deadline. Like others, normalization of relative deadline is also required to generate uniformity for analysis. Equation 7.7 denotes normalized task relative deadline (NTRD).

$$NRD = \frac{Relative\ Deadline\ of\ a\ Non\text{-}Periodic\ Task}{Total\ time\ in\ a\ Schedule\ Period} \quad (7.7)$$

Figure 7.12 depicts how TRR varies against NTRD and NPB. TRR is low if task deadline is low. This is because tasks with low relative deadlines are easy to accommodate in slack that are small. But TRR inversely varies with NPB. This is

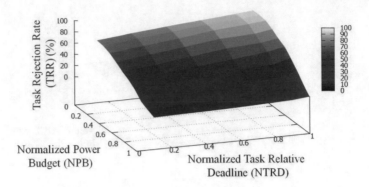

Fig. 7.12 TRR analysis against NTRD and NPB

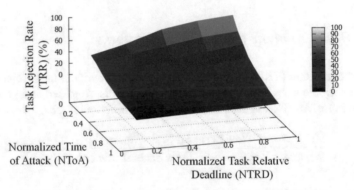

Fig. 7.13 TRR analysis against NTRD and NToA

because if power budget is more, more non-periodic tasks can be executed with enhanced frequency and hence, lower will be TRR.

As per Fig. 7.12, variation of TRR is convex in nature against NTRD and NPB. This is because non-periodic tasks execution takes place on the remaining power budget, i.e. power budget left after allocation of power to periodic tasks. If the power budget is considered static for periodic tasks, then with more power, more non-periodic tasks can execute.

TRR Analysis over NTRD and NToA Fig. 7.13 depicts how TRR varies with respect to NTRD and NToA.

Same as the previous scenario, TRR is low when task relative deadline is low. This is because in small slack spaces, tasks whose relative deadlines are low is possible to accommodate. But TRR inversely varies with NToA. This is because TRR increases if ToA is early, and vice-versa. Moreover, as more slack spaces are present in late phases of a schedule, hence, non-periodic tasks can be re-executed on different resources during an attack.

Graphical analysis indicates a concave nature nature of variation for TRR. This is because if power budget is considered static and HTH attack occurs in initial phases of a schedule, then extra power of power budget is used for re-execution of periodic tasks. Thus the remaining power budget will be low, which is used for execution of non-periodic tasks and thus, the concave nature.

7.7 Conclusion

FPGAs find wide use in various applications in the present era, specifically after the advent of Industry 4.0. But existing energy or power aware real time task scheduling strategies on FPGAs do not consider HTH attacks. Power draining attacks of HTH may decrease the lifetime of the system. In this chapter, we show how power draining attacks may affect real time schedules and propose mechanisms that can ensure completion of periodic task schedules in the midst of such attacks. Along with this, we also focus on handling of non-periodic tasks in slacks of periodic task schedules, based on the remaining power budget. Experimentation is performed with standard FPGAs and schedules are generated with tasks of EPFL benchmark suite. Analysis of results for periodic and non-periodic tasks are performed based on metrics TSR and TRR respectively.

References

[AGDM15] L. Amarú, P.-E. Gaillardon, G. De Micheli,. The EPFL combinational benchmark suite, in *Proceedings of the 24th International Workshop on Logic & Synthesis (IWLS)*, number CONF (2015)

[BC18] K. Baital, A. Chakrabarti, Dynamic scheduling of tasks for multi-core real-time systems based on optimum energy and throughput. IET Comput. Digit. Tech. **13**(2), 93–100 (2018)

[BHBN14] S. Bhunia, M.S. Hsiao, M. Banga, S. Narasimhan, Hardware Trojan attacks: threat analysis and countermeasures. Proc. IEEE **102**(8), 1229–1247 (2014)

[BM13] C. Bolchini, A. Miele, Reliability-driven system-level synthesis for mixed-critical embedded systems. IEEE Trans. Comput. **62**(12), 2489–2502 (2013)

[BMAB16] M. Bambagini, M. Marinoni, H. Aydin, G. Buttazzo, Energy-aware scheduling for real-time systems: a survey. ACM Trans. Embed. Comput. Syst. **15**(1) (2016)

[Boa05] Defense Science Board, Task Force on High Performance Microchip Supply (2005), http://www.acq.osd.mil/dsb/reports/ADA435563.pdf. Accessed 2005

[CMSW14] X. Cui, K. Ma, L. Shi, K. Wu, High-level synthesis for run-time hardware Trojan detection and recovery, in *Proceedings of the 51st Annual Design Automation Conference, DAC'14*, (2014), pp. 157:1–157:6

[GMSC18] K. Guha, A. Majumder, D. Saha, A. Chakrabarti, Reliability driven mixed critical tasks processing on FPGAs against hardware Trojan attacks, in *21st Euromicro Conference on Digital System Design, DSD 2018, Prague, Czech Republic, August 29-31, 2018*, ed. by M. Novotný, N. Konofaos, A. Skavhaug (IEEE Computer Society, 2018), pp. 537–544

[GMSC19]   K. Guha, A. Majumder, D. Saha, A. Chakrabarti, Criticality based reliability against hardware trojan attacks for processing of tasks on reconfigurable hardware. Microprocess. Microsyst. **71** (2019)

[GSC17]   K. Guha, D. Saha, A. Chakrabarti, Real-time SoC security against passive threats using Crypsis behavior of geckos. J. Emerg. Technol. Comput. Syst. **13**(3), 41:1–41:26 (2017)

[GSC18]   K. Guha, A. Majumder, D. Saha, A. Chakrabarti, Criticality based reliability against hardware trojan attacks for processing of tasks on reconfigurable hardware. Microprocess. Microsyst. **71** (2019)

[GSC19a]   K. Guha, D. Saha, A. Chakrabarti, SARP: self aware runtime protection against integrity attacks of hardware Trojans, in *VLSI Design and Test* (Singapore, 2019), pp. 198–209

[GSC19b]   K. Guha, D. Saha, and A. Chakrabarti. Stigmergy-Based Security for SoC Operations From Runtime Performance Degradation of SoC Components. ACM Trans. Embed. Comput. Syst. **18**(2), 14:1–14:26 (2019)

[HKM+14]   T. Hayashi, A. Kojima, T. Miyazaki, N. Oda, K. Wakita, T. Furusawa, Application of FPGA to nuclear power plant i&c systems, in *Progress of Nuclear Safety for Symbiosis and Sustainability* (Springer, 2014), pp. 41–47

[KM18]   H. Koc, P.P. Madupu, Optimizing energy consumption in cyber physical systems using multiple operating modes, in *2018 IEEE 8th Annual Computing and Communication Workshop and Conference (CCWC)* (2018), pp. 520–525

[LJM13]   Y. Liu, Y. Jin, Y. Makris, Hardware Trojans in wireless cryptographic ICs: silicon demonstration & detection method evaluation, in *Proceedings of the International Conference on Computer-Aided Design, ICCAD'13* (2013), pp. 399–404

[LRYK14]   C. Liu, J. Rajendran, C. Yang, R. Karri, Shielding heterogeneous MPSoCs from untrustworthy 3PIPs through security-driven task scheduling. IEEE Trans. Emerg. Topics Comput. **2**(4), 461–472 (2014)

[MDS19]   S. Moulik, R. Devaraj, A. Sarkar, Healers: a heterogeneous energy-aware low-overhead real-time scheduler. IET Comput. Digit. Tech. **13**(6), 470–480 (2019)

[MFP16]   V. Moghaddas, M. Fazeli, A. Patooghy, Reliability-oriented scheduling for static-priority real-time tasks in standby-sparing systems. Microprocess. Microsyst. **45**(PA), 208–215 (2016)

[MKN+16]   S. Mal-Sarkar, R. Karam, S. Narasimhan, A. Ghosh, A. Krishna, S. Bhunia, Design and validation for FPGA trust under hardware Trojan attacks. IEEE Trans. Multi-Scale Comput. Syst. **2**(3), 186–198 (2016)

[MWP+09]   D. McIntyre, F. Wolff, C. Papachristou, S. Bhunia, D. Weyer, Dynamic evaluation of hardware trust, in *2009 IEEE International Workshop on Hardware-Oriented Security and Trust* (2009), pp. 108–111

[MYAH19]   B.J. Mohd, K.M.A. Yousef, A. AlMajali, T. Hayajneh. Power-aware adaptive encryption. in *2019 IEEE Jordan International Joint Conference on Electrical Engineering and Information Technology (JEEIT)* (2019), pp. 711–716

[RB18]   L. Ribeiro, M. Björkman, Transitioning from standard automation solutions to cyber-physical production systems: an assessment of critical conceptual and technical challenges. IEEE Syst. J. **12**(4), 3816–3827 (2018)

[RSK16]   J.J. Rajendran, O. Sinanoglu, R. Karri, Building trustworthy systems using untrusted components: a high-level synthesis approach. IEEE Trans. Very Large Scale Integr. (VLSI) Syst. **24**(9), 2946–2959 (2016)

[SP18]   S.Z. Sheikh, M.A. Pasha, Energy-efficient multicore scheduling for hard real-time systems: a survey. ACM Trans. Embed. Comput. Syst. **17**(6) (2018)

[TK10]   M. Tehranipoor, F. Koushanfar, A survey of hardware Trojan taxonomy and detection. IEEE Des. Test Comput. **27**(1), 10–25 (2010)

[Tos12]   S. Tosun, Energy and reliability aware task scheduling onto heterogeneous MPSoC architectures. Springer J. Supercomput. **62**, 265–289 (2012)

[XFT14]   K. Xiao, D. Forte, M. Tehranipoor, A novel built-in self-authentication technique to prevent inserting hardware Trojans. IEEE Trans. Comput. Aided Des. Integr. Circuits Syst. **33**(12), 1778–1791 (2014)

[XGS+15]   S. Xin, Q. Guo, H. Sun, B. Zhang, J. Wang, C. Chen, Cyber-physical modeling and cyber-contingency assessment of hierarchical control systems. IEEE Trans. Smart Grid **6**(5), 2375–2385 (2015)

[Xil10]   Incorporation Xilinx, Virtex-4 family overview, in *Tech. Doc. DS112 (v2. 0)* (2010), pp. 1–8

[Xil18]   Xilinx_Inc., ZynQ-7000 SoC-Technical Reference Manual UG585, https://www.xilinx.com/support/documentation/user_guides/ug585-Zynq-7000-TRM.pdf (2018)

Correction to: Scheduling Algorithms for Reconfigurable Systems

Correction to:
Chapter 3 in: K. Guha et al., *Self Aware Security for Real Time Task Schedules in Reconfigurable Hardware Platforms,*
https://doi.org/10.1007/978-3-030-79701-0_3

In the original version of the book, the following belated corrections have been incorporated: In Ref. [MVGR13] of Chapter 3, the first author's name has been corrected from "A. Medina Villanueva" to "A. Morales-Villanueva" and the second author's name from "A.G. Ross" to "A. Gordon-Ross". The book and the chapter have been updated with the changes.

The updated version of this chapter can be found at
https://doi.org/10.1007/978-3-030-79701-0_3

Printed in the United States
by Baker & Taylor Publisher Services